Lecture Notes in Mathematics

Edited by A. Dold and B. Eckmann

1065

Annie Cuyt

Padé Approximants for Operators: Theory and Applications

Springer-Verlag
Berlin Heidelberg New York Tokyo 1984

Author

Annie Cuyt
Department of Mathematics UIA, University of Antwerp
Universiteitsplein 1, 2610 Wilrijk, Belgium

AMS Subject Classification (1980): 41A21

ISBN 3-540-13342-9 Springer-Verlag Berlin Heidelberg New York Tokyo
ISBN 0-387-13342-9 Springer-Verlag New York Heidelberg Berlin Tokyo

CIP-Kurztitelaufnahme der Deutschen Bibliothek. Cuyt, Annie: Padé approximants for operators: theory and applications / Annie Cuyt. – Berlin; Heidelberg; New York; Tokyo: Springer, 1984.
(Lecture notes in mathematics; 1065)
ISBN 3-540-13342-9 (Berlin ...);
ISBN 0-387-13342-9 (New York ...)
NE: GT

Printing and binding: Beltz Offsetdruck, Hemsbach/Bergstr.
2146/3140-543210

To my darling husband.

ACKNOWLEDGEMENTS

I hereby want to express my gratitude to Prof. Dr. L. Wuytack (University of Antwerp) for the valuable discussions we had, and to Prof. Dr. H. Werner (University of Bonn) for the interesting comments he gave after reading the manuscript. I may not forget Prof. Dr. Louis B. Rall (University of Wisconsin) for his book gave me the necessary background to start working.

I also thank the Department of Mathematics of the University of Antwerp for their hospitality, the secretaries of the department for the careful typing of all the results, miss B. Verdonk for correcting the final typographical errors, the I.W.O.N.L. (Instituut voor aanmoediging van het Wetenschappelijk Onderzoek in Nijverheid en Landbouw) and N.F.W.O. (Nationaal Fonds voor Wetenschappelijk Onderzoek) for their financial support.

Last but not least I thank my husband for the encouragement.

TABLE OF CONTENTS

LIST OF NOTATIONS

	Significance
X	11
Y	11
O	2
I	2
Λ	1
$\otimes$	4
$F, G, \ldots$	nonlinear operators
$L(X^k, Y)$	3
$P, Q, R, S, T, U, V, W, \ldots$	abstract polynomials
∂P	6
$\partial_0 P$	6
$F^{(k)}(x_0), P^{(k)}(x_0), \ldots$	6
$D(F)$	13
$B(x_0, r)$	5
$O(x^j)$	13
$C_k x^k$	15
$\frac{1}{Q_\star} \cdot P_\star$	13
$\sim$	14
$P_{[n,m]}(x)$	17
$Q_{[n,m]}(x)$	17
$F_i(x)$	17
$\overline{F}_i(x)$	17

SUMMARY

In chapter I the concept of Padé approximants is generalized for nonlinear operators $F: X \to Y$ where X is a Banach space and Y is a commutative Banach algebra, starting from analyticity as is done in the classical theory. The generalization is such that the classical univariate Padé approximant ($X = \mathbb{R} = Y$) is a special case of the theory. We discuss the existence and unicity of a solution of the Padé-approximation problem of order (n,m) for F and prove that a lot of the properties for univariate Padé approximants remain valid: several covariance properties, recurrence relations, the epsilon algorithm , the qd-algorithm, the structure of the Padé table, criteria for regularity and normality of an entry of the Padé table. We are also able to prove a projection property and a product property.
In chapter II the multivariate Padé approximants ($X = \mathbb{R}^p$, $Y = \mathbb{R}$) are studied more extensively. We prove for instance the nontriviality of a solution of the Padé-approximation problem and the near-Toeplitz structure of the homogeneous system of equations. Also an extra covariance property and more recurrence relations are formulated. The multivariate Padé approximants introduced here are compared with other definitions of Padé approximants for multivariate functions given by different authors in the last few years. Our definition turns out to be an interesting generalization too.
Most of the applications are discussed in chapter III, except the acceleration of convergence of a table with multiple entry which is done by means of multivariate Padé approximants and therefore added to chapter II.
As far as the nonlinear operator equations are concerned, we treat the solution of nonlinear systems of equations, initial value problems, boundary value problems, partial differential equations and integral equations. An interesting procedure, especially in the neighbourhood of singularities, is the Halley-iteration which is newly introduced here. Its numerical stability for the solution of a system of nonlinear equations is formulated at the end of chapter III.

CHAPTER I: ABSTRACT PADE APPROXIMANTS IN OPERATOR THEORY

§ 1. MOTIVATION

Padé approximants are a frequently used tool for the solution of mathematical problems: the solution of a nonlinear equation, the acceleration of convergence, numerical integration by means of nonlinear techniques, the solution of ordinary and partial differential equations. In the neighbourhood of singularities the use of Padé approximants can be very interesting.
Many attempts have been made to generalize the concept of Padé approximants in some sense; we refer to definitions of multivariate Padé approximants by Bose [7], Chisholm [11, 12, 13, 14], Karlsson and Wallin [32], Levin [34] and Lutterodt [37], to quadratic approximants and their generalizations [45, 21], to operator Padé approximants for formal power series in a parameter with non-commuting elements of a certain algebra as coefficients [4], to matrix-valued Padé approximants [3, 46], to Padé approximants for the operator exponential [17] and so on.

It would be important to generalize the concept of Padé approximants for nonlinear operators, following the ideas of the classical theory, for this would enable us to prove a lot of the classical properties for the generalizations as well and it would also enable us to use those generalizations for the solution of nonlinear operator equations. these are more general problems than the ones we solved with the aid of univariate Padé approximants; we mention nonlinear systems of equations, nonlinear initial value and boundary value problems, nonlinear partial differential equations and nonlinear integral equations.
Such a generalization is treated here.

§ 2. INTRODUCTION

2.1. *Banach spaces and Banach algebras*

In ordinary analysis we work with the real or complex number system. Here we shall work in complete normed spaces which are generalizations of these number systems. Since linear spaces may consist of such interesting mathematical objects as vectors with a finite or infinite number of components or functions that satisfy given conditions, we shall be able to deal with a wide variety of problems.
In abstract terms, a linear vector space X over the scalar field Λ (where Λ is $\mathbb{R}$ or $\mathbb{C}$) is a set of elements with two operations, called addition and scalar multiplication, which satisfy certain conditions:

a) the set X is a commutative group with respect to the operation of addition (we shall denote the unit for the addition by 0)

b) for any scalars λ, μ in Λ and any elements x, y in X, the following rules hold:

$$\lambda x \in X$$
$$1.x = x$$
$$o.x = 0$$
$$(\lambda+\mu)x = \lambda x+\mu x$$
$$\lambda(x+y) = \lambda x+\lambda y$$

The algebraic structure of a linear space is similar to that of the real or complex number system. However, to deal with other concepts of theoretical and computational importance, such as accuracy of approximation, convergence of sequences, and so on, it is necessary to introduce additional structure into such spaces.

X is called a normed linear space if for each element x in X, a finite non-negative real number $\|x\|$, called the norm of x, is defined and the following conditions are satisfied:

a) $\|x\| = o$ if and only if $x = 0$

b) $\|\lambda x\| = |\lambda| \; \|x\|$

c) $\|x+y\| \leq \|x\| + \|y\|$

In the solution of many problems the basic issue is the existence of a limit x^* of an infinite sequence $\{x_i\}$ of elements of X. A normed linear space X is said to be complete if every Cauchy sequence of elements of X converges to a limit which is an element of X. Such a complete normed linear space is called a Banach space.

Some Banach spaces have the property that the product xy of two elements of the space is defined and is also an element of the space. Such a Banach space is called a Banach algebra if

$$\|xy\| \leq \|x\|.\|y\|$$

A Banach algebra is said to be commutative if

$$xy = yx$$

and we say that it has a unit for the multiplication, which we shall denote by I, if

$$x.I = x = I.x$$

The spaces $\mathbb{R}^p$ and $\mathbb{C}^p$ for example are Banach algebras with unit if the multiplication is defined component-wise.

2.2. *Linear and multilinear operators*

Many mathematical operations which transform one vector or function into another have certain simple algebraic properties. We shall now discuss such operators.

An operator L which maps a linear space X into a linear space Y over the same scalar field Λ so that for each x in X there is a uniquely defined element Lx in Y, is called linear if it is

a) additive: $L(x_1+x_2) = Lx_1 + Lx_2$

b) homogeneous: $L(\lambda x) = \lambda Lx$

If $X=\mathbb{R}^p$ and $Y=\mathbb{R}^q$ then a linear operator L has a unique representation as a qxp matrix. Another example of a linear operator is furnished by differentiation; the operator $D=\frac{d}{dt}$ maps $X=C'([0,1])$ into $Y=C([0,1])$ with

$$Dx(t) = \frac{dx}{dt} = y(t)$$

If X and Y are linear spaces over a common scalar field Λ, then the set of all linear operators from X into Y becomes a linear space over Λ if addition is defined by

$$(L_1+L_2)x = L_1x+L_2x$$

and scalar multiplication by

$$(\lambda L)x = \lambda(Lx)$$

The norm of a linear operator L is defined by

$$\|L\| = \sup_{\|x\|=1} \|Lx\|$$

and the operator L is called bounded if $\|L\| < \infty$.

We know that a continuous linear operator L from a Banach space X into a Banach space Y is bounded [41 pp. 38] and also that

$$\|Lx\| \leq \|L\|.\|x\|$$

Clearly the set $L(X,Y)$ of all bounded linear operators from a Banach space X into a Banach space Y is a Banach space itself. So we may consider linear operators which map X into $L(X,Y)$. For such an operator B and for x_1 and x_2 in X, we would have

$$Bx_1 = L$$

a linear operator from X into Y, and

$$Bx_1x_2 = (Bx_1)x_2$$

an element of Y.

The operator B is called a bilinear operator from X into Y. Since the bounded linear operators from X into $L(X,Y)$ form themselves a linear space $L(X,L(X,Y))$ which we shall denote by $L(X^2,Y)$, the foregoing process could be repeated, leading to a whole hierarchy of linear operators and spaces. These classes of operators play a fundamental role in the differential calculus in Banach spaces.

A k-linear operator L on X is an operator $L:X^k \to Y$ which is linear and homogeneous in each of its arguments separately. If $x_1=\ldots=x_k=x$, we shall use the notation

$$Lx^k = Lx_1\ldots x_k$$

We write $L(X^k,Y)$ for the set of all bounded k-linear operators from X into Y.

We define a o-linear operator on X to be a constant function, i.e. for y fixed in Y, we have

$$Lx = y \text{ for all } x \text{ in } X$$

The set $L(X^o,Y)$ is identified with Y.

If $L \in L(X^k,Y)$ and $x_1,\ldots,x_\ell \in X$ with $k \geq \ell \geq 1$ then

$$Lx_1\ldots x_\ell$$

is a bounded $(k-\ell)$-linear operator.

In general the elements $Lx_1 \dots x_k$ and $Lx_{i_1} \dots x_{i_k}$ with $(x_1,\dots,x_k)$ in X^k and $(i_1,\dots,i_k)$ a permutation of $(1,\dots,k)$ are different, so that actually k! k-linear operators are associated with a given k-linear operator L.
But if

$$L\, x_1 \dots x_k = L\, x_{i_1} \dots x_{i_k}$$

for all $(x_1,\dots,x_k)$ in X^k and for all permutations $(i_1,\dots,i_k)$ of $(1,\dots,k)$ [41 pp. 103-104] then the k-linear bounded operator L is called symmetric.
If Y is a Banach algebra, multilinear operators can also be obtained by forming tensor products.

Definition I.2.1.:

Let $F : X \to Y$ and $G : X \to Y$ be operators.

The product F.G is defined by $(F.G)(x) = F(x).G(x)$ in Y.

Definition I.2.2.:

Let $X_1, \dots, X_p, Z_1, \dots, Z_q$ be vector spaces and let

$F : X_1 \times \dots \times X_p \to Y$ be bounded and p-linear and

$G : Z_1 \times \dots \times Z_q \to Y$ be bounded and q-linear.

The tensorproduct $F \otimes G : X_1 \times \dots \times X_p \times Z_1 \times \dots \times Z_q \to Y$ is bounded and (p+q)-linear when defined by

$$(F \otimes G)\, x_1 \dots x_p\, z_1 \dots z_q = F\, x_1 \dots x_p \,.\, G z_1 \dots z_q$$

[23 pp. 318].

2.3. *Fréchet-derivatives*

An operator F from X into Y is called nonlinear if it is not a linear operator.
Now suppose that F is an operator that maps a Banach space X into a Banach space Y.
If L in L(X,Y) exists such that

$$\lim_{\|\Delta x\| \to o} \frac{\| F(x_o+\Delta x) - F(x_o) - L\Delta x \|}{\|\Delta x\|} = o$$

then F is said to be Fréchet-differentiable at x_o, and the bounded linear operator

$$L = F'(x_o)$$

is called the first Fréchet-derivative of F at x_o.
Note that the classical rules for differentiation, like the chain rule still hold for Fréchet differentiation. In practice, to differentiate a given nonlinear operator F, we attempt to write the difference $F(x_o+\Delta x) - F(x_o)$ in the form

$$F(x_o+\Delta x) - F(x_o) = L(x_o,\Delta x)\Delta x + \eta(x_o,\Delta x)$$

where $L(x_0,\Delta x)$ is a bounded linear operator for given x_0 and Δx with

$$\lim_{\|\Delta x\|\to 0} L(x_0,\Delta x) = L \in L(X,Y)$$

and

$$\lim_{\|\Delta x\|\to 0} \frac{\|\eta(x_0,\Delta x)\|}{\|\Delta x\|} = 0$$

To illustrate this process, consider the operator F in $C([0,1])$ defined by

$$F(x) = x(t)\int_0^1 \frac{t}{s+t}\, x(s)\, ds \qquad 0 \le t \le 1$$

The difference $F(x_0+\Delta x) - F(x_0)$ equals

$$x_0(t)\int_0^1 \frac{t}{s+t}\,\Delta x(s)\, ds + \Delta x(t)\int_0^1 \frac{t}{s+t}\, x_0(s)\, ds + \Delta x(t)\int_0^1 \frac{t}{s+t}\,\Delta x(s)\, ds$$

So the operator $L(x_0,\Delta x)$ equals

$$x_0(t)\int_0^1 \frac{t}{s+t}[\ \ ]\, ds + [\ \ ]\int_0^1 \frac{t}{s+t}\, x_0(s)\, ds + [\ \ ]\int_0^1 \frac{t}{s+t}\,\Delta x(s)\, ds$$

where [ ] is a place holder and is used to indicate the position of the argument of the operator $L(x_0,\Delta x)$.

Now $L(x_0,\Delta x)$ is a continuous function of Δx; so we may set $\Delta x = 0$ to obtain $F'(x_0) = L$:

$$F'(x_0) = x_0(t)\int_0^1 \frac{t}{s+t}[\ \ ]\, ds + [\ \ ]\int_0^1 \frac{t}{s+t}\, x_0(s)\, ds$$

where now [ ] indicates the position of the argument of the linear operator $F'(x_0)$.

Suppose that an operator F from X into Y is differentiable at x_0 and also at every point of the open ball $B(x_0,r)$ with centre x_0 and radius $r > 0$. For each x in $B(x_0,r)$ $F'(x_0)$ will be an element of the space $L(X,Y)$. Consequently F' may be considered to be an operator defined in a neighbourhood of x_0. We know that F' will be differentiable at x_0 if a bounded linear operator B from X into $L(X,Y)$ exists such that

$$\lim_{\|\Delta x\|\to 0} \frac{\|F'(x_0+\Delta x) - F'(x_0) - B\Delta x\|}{\|\Delta x\|} = 0$$

Such a bounded linear operator B is known to be a bilinear operator and if it exists, it is called the second derivative of F at x_0 and denoted by $F''(x_0) = B$. Thus the second derivative of an operator F is obtained by differentiating its first derivative F'. Now it is possible to give an inductive definition of higher derivatives of an operator F.

2.4. *Abstract polynomials*

If L is a k-linear operator from a Banach space X into a Banach algebra Y, then the operator P from X into Y defined by

$$P(x) = Lx^k \text{ for } x \text{ in } X$$

is a nonlinear operator. In this way we can define abstract polynomials.

Definition I.2.3.:

An <u>abstract polynomial</u> is a nonlinear operator $P : X \to Y$ such that
$P(x) = A_n x^n + A_{n-1} x^{n-1} + \dots + A_o$ with
$A_i \in L(X^i, Y)$ and A_i symmetric [41 p. 107].

The degree of $P(x)$ is n. We also introduce the following notations.
If there exists a positive integer j_1 such that for all $o \leq k < j_1$: $A_k x^k \equiv 0$ and $A_{j_1} x^{j_1} \not\equiv 0$ then $\partial_o P = j_1$ is called the order of the abstract polynomial P.
If there exists a positive integer j_2 such that for all $j_2 < k \leq n$: $A_k x^k \equiv 0$ and $A_{j_2} x^{j_2} \not\equiv 0$ then $\partial P = j_2$ is called the exact degree of the abstract polynomial P.
Abstract polynomials are differentiated as in elementary calculus: if $P(x) = A_n x^n + A_{n-1} x^{n-1} + \dots + A_o$ then the Fréchet-derivatives of P at x_o are

$$P'(x_o) = n A_n x_o^{n-1} + \dots + 2 A_2 x_o + A_1 \in L(X, Y)$$

$$P^{(2)}(x_o) = n(n-1) A_n x_o^{n-2} + \dots + 2 A_2 \in L(X^2, Y)$$

$$\vdots$$

$$P^{(n)}(x_o) = n!\, A_n \in L(X^n, Y)$$

We emphasis the fact that for an operator $F: X \to Y$, the kth Fréchet-derivative at x_o, $F^{(k)}(x_o)$, is a symmetric k-linear and bounded operator [41 pp. 110]. Examples of abstract polynomials and k^{th} Fréchet-derivatives of a nonlinear operator can be found in § 3. of this chapter.
We can easily prove the following important lemmas for abstract polynomials.

Lemma I.2.1.:

Let the abstract polynomial P be given by $P(x) = \sum_{i=o}^{n} A_i x^i$.
If $P(x) \equiv 0$ then $A_i = 0$ for $i = o, \dots, n$.

Lemma I.2.2.:

Let V be an abstract polynomial and U a continuous operator with $D(U) \neq \emptyset$.
If $U(x).V(x) \equiv 0$ then $V(x) \equiv 0$.

Proof:

Since $D(U) \neq \emptyset$, we can find x_o in X such that $U(x_o)$ is regular.

For the abstract polynomial V(x) we can write

$$V(x) = \sum_{k=o}^{n} \frac{1}{k!} V^{(k)} (x_0) (x - x_0)^k \qquad [41 \text{ pp. } 111]$$

$$\text{with } \frac{1}{o!} V^{(o)} (x_0) (x - x_0)^0 = V(x_0)$$

Now $U(x_0).V(x_0) = 0$ and so $V(x_0) = 0$.

Since U is continuous, D(U) is an open set. Thus there is an open ball $B(x_0, r)$ with centre x_0 and radius $r > o$, such that $B(x_0, r) \subset D(U)$, in other words such that for all x in $B(x_0, r)$: U(x) is regular.

This implies that for all x in $B(x_0, r) : V(x) = 0$.

For $V'(x_0)$ we can write

$$\lim_{\|h\| \to o} \frac{\|V(x_0+h) - V(x_0) - V'(x_0)h\|}{\|h\|} = o$$

Or equivalently for $\|h\| < r$

$$\lim_{\|h\| \to o} \frac{\|V'(x_0)h\|}{\|h\|} = o$$

So $\forall \varepsilon > o, \exists \delta > o: \|h\| < \gamma = \min(\delta, r) \Rightarrow \|V'(x_0)h\| \le \varepsilon\|h\|$

Take x in $X\backslash\{0\}$. Then $\|\frac{1}{2}\gamma \; \|x\|^{-1} \; x\| < \gamma$ and so

$$\|V'(x_0)(\tfrac{1}{2}\gamma \; \|x\|^{-1} \; x)\| \le \varepsilon\|\tfrac{1}{2}\gamma \; \|x\|^{-1} \; x\|$$

or equivalently, since $\frac{\gamma}{2\|x\|} > o$

$$\|V'(x_0) \; x\| \le \varepsilon\|x\| \qquad (I.2.1)$$

For $x = 0$ also $\|V'(x_0) \; x\| \le \varepsilon\|x\|$

Now $\|V'(x_0)\| = \inf\{M \ge o \mid \|V'(x_0) \; x\| \le M\|x\| \quad \text{for all x in X}\}$

and thus (I.2.1) implies $\|V'(x_o)\| = o$.

For every x in X we have now $\|V'(x_o)\, x\| \leq \|V'(x_o)\|.\|x\| = o$

and so $V'(x_o)\, x \equiv 0$ or $V'(x_o) = 0$ as operator $X \to Y$.

To proceed, take x in $B(x_o,r)$. A radius $r_o > o$ exists such that

for every y in $B(x, r_o) \subset B(x_o,r) : V(y) = 0$

So we can prove that for all x in $B(x_o, r) : V'(x) = 0$.

Repeating the previous procedure,

we can now prove that $V^{(2)}\ (x_o) = 0$.

And so on till we have $V^{(n)}\ (x_o) = 0$ and thus $V(x) \equiv 0$.

∎

Lemma I.2.3.:

Let the nontrivial abstract polynomials V and W be given by $V(x) = \sum_{i=v_1}^{v_2} V_i x^i$ and $W(x) = \sum_{j=w_1}^{w_2} W_j x^j$ with $\partial_o V = v_1$ and $\partial V = v_2$, $\partial_o = w_1$ and $\partial W = w_2$.

If $D(V) \neq \emptyset$ and Y is a commutative Banach algebra without nilpotent elements, then $\partial W \leq \partial(V.W) - v_1$

Proof:

If V(x) or W(x) are monomials, the proof is trivial.

Write $\partial_o(V.W) = p_1$ and $\partial(V.W) = p_2$; we always have that

$v_1+w_1 \leq p_1 \leq p_2 \leq v_2 + w_2$.

Suppose $\partial W \geq \partial(V.W) - v_1 + 1$.

$$\text{Then}\ \left\{\begin{array}{l} V_{v_2}\, x^{v_2} . W_{w_2}\, x^{w_2} \equiv 0 \\ \\ V_{v_2}\, x^{v_2} . W_{w_2-1}\, x^{w_2-1} + V_{v_2-1}\, x^{v_2-1} . W_{w_2}\, x^{w_2} \equiv 0 \\ \vdots \\ V_{p_2+1-w_1}\, x^{p_2+1-w_1} . W_{w_1}\, x^{w_1} + \ldots + V_{v_1}\, x^{v_1} . W_{p_2+1-v_1}\, x^{p_2+1-v_1} \equiv 0 \end{array}\right.$$

with $p_2+1 \leq w_2+v_1$.

This implies

$$\begin{cases} V_{v_2} x^{v_2} \, . \, W_{w_2} x^{w_2} \equiv 0 \\ V_{v_2-1} x^{v_2-1} \, . \, (W_{w_2} x^{w_2})^2 \equiv 0 \\ \vdots \\ V_{v_1} x^{v_1} \, . \, (W_{w_2} x^{w_2})^{1+v_2-v_1} \equiv 0 \end{cases}$$

and thus $V(x) \, . \, (W_{w_2} x^{w_2})^{1+v_2-v_1} \equiv 0$

Since Y contains no nilpotent elements: $W_{w_2} x^{w_2} \equiv 0$.

This contradicts $\partial W = w_2$.

■

The following example will illustrate that for lemma I.2.3 we really need a commutative Banach algebra Y without nilpotent elements.

Consider $Y = \left\{ \begin{pmatrix} a_1 & 0 & 0 \\ a_2 & a_1 & 0 \\ a_3 & a_2 & a_1 \end{pmatrix} \middle| \; a_1, a_2, a_3 \in \mathbb{R} \right\}$ normed by

$$\left\| \begin{pmatrix} a_1 & 0 & 0 \\ a_2 & a_1 & 0 \\ a_3 & a_2 & a_1 \end{pmatrix} \right\| = 3 \max(|a_1|, |a_2|, |a_3|).$$

It is easy to verify that Y is a commutative Banach algebra.

Let $X = \mathbb{R}$.

Take $V(x) = \begin{pmatrix} -1 & 0 & 0 \\ 0 & -1 & 0 \\ 0 & 0 & -1 \end{pmatrix} + \begin{pmatrix} 0 & 0 & 0 \\ x & 0 & 0 \\ 0 & x & 0 \end{pmatrix}$

and $W(x) = \begin{pmatrix} 0 & 0 & 0 \\ 1 & 0 & 0 \\ 0 & 1 & 0 \end{pmatrix} + \begin{pmatrix} 0 & 0 & 0 \\ 0 & 0 & 0 \\ x & 0 & 0 \end{pmatrix}$.

So $\partial_o V = o = \partial_o W$, $\partial V = 1 = \partial W$, $0 \in D(V)$ and $V_1 x$ and $W_1 x$ are nilpotent elements.

$$\text{Now } (V.W)(x) = \begin{pmatrix} 0 & 0 & 0 \\ -1 & 0 & 0 \\ 0 & -1 & 0 \end{pmatrix} .$$

Clearly $\partial W = 1 > \partial(V.W) - \partial_o V = o$.

3. DEFINITION

1. *Univariate Padé approximant*

Let us first briefly repeat the definition of Padé approximant for a real-valued function F of one real variable, given by its Taylor series development in the origin:

$$F(x) = \sum_{k=o}^{\infty} c_k x^k$$

with

$$c_k = \frac{1}{k!} F^{(k)}(0)$$

First choose n and m in $\mathbb{N}$. Then find two polynomials $P(x) = \sum_{i=o}^{n} a_i x^i$ and $Q(x) = \sum_{j=o}^{m} b_j x^j$ satisfying

$$\partial_o (F.Q-P) \geq n+m+1$$

In other words the $a_o,\dots,a_n,b_o,\dots b_m$ satisfy the following systems of equations:

$$\begin{cases} c_o b_o = a_o \\ c_1 b_o + c_o b_1 = a_1 \\ \vdots \\ c_n b_o + c_o b_n = a_n \end{cases}$$

$$\begin{cases} c_{n+1} b_o + \dots + c_{n+1-m} b_m = o \\ \vdots \\ c_{n+m} b_o + \dots + c_n b_m = o \end{cases}$$

with b_j=o for $j > m$ and c_k=o for $k < o$.

It is obvious that the homogeneous system always has a nontrivial solution since one of the b_j can be chosen freely; the solution for the b_j can then be substituted in the system of equations that gives the coefficients a_i.

After calculation of the polynomials P(x) and Q(x) the (n,m) Padé approximant is defined as the irreducible form of the rational function $\frac{P(x)}{Q(x)}$. The long history of Padé approximants has extensively been studied by Brezinski and their properties have very nicely been formulated and discussed by Baker [2]. The interested reader is referred to their books. We will treat here the generalization to the operator case of the definition and all the properties.
From now on, let X be a Banach space and Y a commutative Banach algebra with unit I for the multiplication.

3.2. *Abstract analyticity*

To generalize the notion of Padé approximant we start from analyticity, as in elementary calculus.

Definition I.3.1.:

The operator $F : X \to Y$ is <u>abstract analytic</u> in x_0 if there exists $B(x_0, r)$ with $r > 0$ such that

$$F(x_0+h) = \sum_{k=0}^{\infty} \frac{1}{k!} F^{(k)} (x_0) h^k \text{ for } \|h\| < r \qquad [41 \text{ pp. } 113]$$

with $\frac{1}{0!} F^{(0)}(x_0)\, h^0 = F(x_0)$.

We give some examples of such series.

a) $C([0,1])$ with the supremum-norm is a commutative Banach algebra if addition and multiplication are performed pointwise.
Consider the Nemyckii-operator $G : C([0,1]) \to C([0,1]) : x \to g(s, x(s))$ with $g \in C^{(\infty)}([0,1] \times C([0,1]))$. Let $I_x : C([0,1]) \to C([0,1]) : x \to x$.

Then clearly $G^{(k)}(x_0) = \frac{\partial^k g}{\partial x^k}(s, x_0(s)).\underbrace{I_x \otimes \ldots \otimes I_x}_{k \text{ times}}$

and so $G(x) = \sum_{k=0}^{\infty} \frac{1}{k!} \frac{\partial^k g}{\partial x^k}(s, x_0(s)).(x-x_0)^k$ in a neighbourhood of x_0 [41 pp. 95].

b) Consider the Urysohn integral operator $U : C([0,1]) \to C([0,1])$:
$x \to \int_0^1 f(s, t, x(t))\, dt$ with $f \in C^{(\infty)}([0,1] \times [0,1] \times C([0,1]))$.
Let [ ] indicate a place-holder for $x(t) \in C([0,1])$.

Then we write $U^{(k)}(x_o) = \int_0^1 \frac{\partial^k f}{\partial x^k}(s, t, x_o(t)) \underbrace{[\,] \ldots [\,]}_{k \text{ times}} dt$

and so $U(x) = \sum_{k=o}^{\infty} \frac{1}{k!} \int_0^1 \frac{\partial^k f}{\partial x^k} (s, t, x_o(t))(x-x_o)^k (t)dt$ in a

neighbourhoud of x_o [41 pp. 97].

c) $C^{(i)}([0,T])$, normed by $\|x(t)\|_\infty = \max\{\|x^{(j)}(t)\| \mid j = o, \ldots, i\}$ or $\|x(t)\|_1 = \sum_{j=o}^{i} \|x^{(j)}(t)\|$ where $\|x^{(j)}(t)\|$ is the norm chosen in $C([0,T])$, is a Banach space.

Consider the operator $V : C'([0,T]) \to C([0,T]) : y \to \frac{dy}{dt} - f(t,y)$

in the initial value problem $V(y) = 0$ with $y(0) = a \in \mathbb{R}$.

Let $I_y : C'([0,T]) \to C([0,T]) : y \to y$.

We see that $V'(y_o) = \frac{d}{dt} - \frac{\partial f(t,y)}{\partial y}(t, y_o) \cdot I_y$ and

$$V^{(k)}(y_o) = - \frac{\partial^k f(t,y)}{\partial y^k}(t, y_o) \cdot \underbrace{I_y \otimes \ldots \otimes I_y}_{k \text{ times}} \text{ for } k \geq 2.$$

So $V(y) = - f(t, y_o) + \frac{dy}{dt} - \sum_{k=1}^{\infty} \frac{1}{k!} \frac{\partial^k f(t,y)}{\partial y^k}(t,y_o) \cdot (y-y_o)^k (t)$

in a neighbourhood of y_o.

d) $\mathbb{R}^p$ and $\mathbb{C}^p$ with componentwise addition and multiplication are commu tative Banach algebras.

Finally let this nonlinear system of 2 real variables be given

$$F\binom{x}{y} = \begin{pmatrix} 1+x+\sin(xy) \\ x^2+y^2-4xy \end{pmatrix}$$

For $x_o = \binom{0}{0}$ we can write

$$F\binom{x}{y} = \binom{1}{0} + \binom{x}{0} + \begin{pmatrix} xy \\ x^2+y^2-4xy \end{pmatrix} + \sum_{k=1}^{\infty} \begin{pmatrix} (-1)^k \frac{(xy)^{2k+1}}{(2k+1)!} \\ 0 \end{pmatrix}$$

3.3. *Abstract Padé approximant*

We call an element y of Y regular if there exists y^{-1} in Y such that $y.y^{-1} = I = y^{-1}.y$ and we write $D(F) = \{x \in X | F(x) \text{ is regular in } Y\}$. The set D(F) is an open set in X if F is continuous [33 pp. 31]. If the operator G maps X into Y, we can define the operator $\frac{1}{G}$ that maps D(G) into Y, by

$$\frac{1}{G}(x) = [G(x)]^{-1}$$

Let now x_0 in definition I.3.1 equal 0 without loss of generality and let F: X → Y be a nonlinear operator abstract analytic in 0:

$$F(x) = \sum_{k=0}^{\infty} \frac{1}{k!} F^{(k)}(0)\, x^k$$

Definition I.3.2.:

The operator $F(x) = O(x^j)$ ($j \in \mathbb{N}$) if there exist nonnegative numbers $s < 1$ and J such that $\|F(x)\| \le J.\|x\|^j$ for all x in B(0,s).

Definition I.3.3.:

The couple of abstract polynomials (P(x), Q(x)) =

$(A_{nm+n}\, x^{nm+n} + \ldots + A_{nm}\, x^{nm},\ B_{nm+m}\, x^{nm+m} + \ldots + B_{nm}\, x^{nm})$

such that the abstract power series $(F.Q-P)(x) = O(x^{nm+n+m+1})$ (I.3.1)

is called a solution of the Padé-approximation problem of order (n,m).

The shift of degrees in P(x) and Q(x) by n.m will be justified in § 4. and § 7. of this chapter. We shall restrict ourselves now to those n and m for which a solution (P(x),Q(x)) with $D(P) \cup D(Q) \neq \emptyset$ can be found.

Definition I.3.4.:

The abstract rational operator $\frac{1}{Q}.P$, the quotient of two abstract polynomials, is reducible if there exist abstract polynomials T, P_* and Q_* such that $P = P_*.T$, $Q = Q_*.T$, $\partial T \ge 1$ and $\frac{1}{T}$ is not an abstract polynomial (i.e. T is not a unit in the ring of abstract polynomials).

Since $D(P) \cup D(Q) \neq \emptyset$, we know that $D(T) \neq \emptyset$ and $D(P_*) \cup D(Q_*) \neq \emptyset$.
For the solutions (P,Q) of the Padé-approximation problem and for the reduced rational operators $\frac{1}{Q_*}.P_*$ we will prove the following equivalence-property.

Theorem I.3.1.:

Let (P,Q) and (R,S) satisfy definition I.3.3, with $D(P) \cup D(Q) \neq \emptyset$ and $D(R) \cup D(S) \neq \emptyset$. Let $\frac{1}{Q_\star}.P_\star$ be a reduced form of the rational operator $\frac{1}{Q}.P$ and $\frac{1}{S_\star}.R_\star$ be a reduced form of the rational operator $\frac{1}{S}.R$.

Then for all x in X: $P(x).S(x) = Q(x).R(x)$

$$P_\star(x).S(x) = Q_\star(x).R(x)$$

$$P_\star(x).S_\star(x) = Q_\star(x).R_\star(x)$$

Proof:

Consider

$P(x).S(x)-R(x).Q(x) = [F(x).S(x)-R(x)].Q(x)-[F(x).Q(x)-P(x)].S(x)$

Now $(F.Q-P)(x) = O(x^{nm+n+m+1})$ and $(F.S-R)(x) = O(x^{nm+n+m+1})$.

The series $[(F.S-R).Q-(F.Q-P).S](x) = O(x^{2nm+n+m+1})$

while $\partial(P.S-R.Q) \leq 2nm+n+m$.

So $(P.S-R.Q)(x) \equiv 0$.

If $P = P_\star.T$ and $Q = Q_\star.T$ with $P_\star$, $Q_\star$ and T abstract polynomials

then $(P.S-R.Q)(x) = [T.(P_\star.S-R.Q_\star)](x)$

Since $D(T) \neq \emptyset$, lemma I.2.2 says that $(P_\star.S-R.Q_\star)(x) \equiv 0$.

If $R = R_\star.U$ and $S = S_\star.U$ with $R_\star$, $S_\star$ and U abstract polynomials

then $(P_\star.S-R.Q_\star)(x) = [U.(P_\star.S_\star-R_\star.Q_\star)](x)$.

Since $D(U) \neq \emptyset$, lemma I.2.2 says that $(P_\star.S_\star-R_\star.Q_\star)(x) \equiv 0$. ∎

We write $A = A_1 \cup A_2$ with

$A_1 = \{(P,Q) \mid (P,Q)$ satisfies definition I.3.3 for a certain n and m and $D(P) \cup D(Q) \neq \emptyset\}$

$A_2 = \{(P_\star,Q_\star) \mid \frac{1}{Q_\star}.P_\star$ is a reduced form of the rational operator $\frac{1}{Q}.P$ for (P,Q) in $A_1\}$

Clearly the relation $(P_1,Q_1) \sim (P_2,Q_2)$ if and only if $P_1(x).Q_2(x) = Q_1(x).P_2(x)$ for all x in X, is an equivalence-relation in A which divides A in disjoint equivalence-classes.

Definition I.3.5.:

The equivalence-class of A containing a solution of (I.3.1) for n and m chosen, will be called the (n,m) abstract Padé-approximant for F.

This equivalence-class does not always contain a couple of abstract polynomials $(P_\star, Q_\star)$ with $Q_\star(0) = I$. An example will illustrate this. Consider the operator

$$F : \mathbb{R}^2 \to \mathbb{R}^2 : \binom{x}{y} \to \begin{pmatrix} 1+x+\sin(xy) \\ x^2+y^2-4xy \end{pmatrix} = \binom{1}{0} + \binom{x}{0} + \begin{pmatrix} xy \\ x^2+y^2-4xy \end{pmatrix} + \ldots$$

Take n = 1 = m. The couple of abstract polynomials

$$(P_\star(x,y),\ Q_\star(x,y)) = \begin{pmatrix} 1+x-y & & 1-y \\ & , & \\ 0 & & 1 \end{pmatrix}$$ belongs to the (n,m) Padé-approximant for F.

Here $Q_\star\binom{0}{0} = \binom{1}{1}$.

Take n = 1 and m = 2. The couple of abstract polynomials

$$(P_\star(x,y),\ Q_\star(x,y)) = \begin{pmatrix} x-y+x^2-2xy & & x-y-xy+xy^2 \\ & , & \\ 0 & & 1 \end{pmatrix}$$ is numerator and

denominator of the irreducible form of $\frac{P(x,y)}{Q(x,y)}$ where $(P(x,y),\ Q(x,y))$ is any nontrivial solution of (I.3.1).

Here $Q_\star\binom{0}{0} = \binom{0}{1}$ because the order of the first component in $Q_\star(x,y)$ is 1 and no further reduction can be performed to lower this order.

§ 4. EXISTENCE OF A SOLUTION

We will now discuss the existence and calculation of a solution of (I.3.1). Write $\frac{1}{k!}F^{(k)}(0) = C_k$, a symmetric k-linear bounded operator. The condition (I.3.1) is equivalent with (I.4.1) and (I.4.2):

$$
(\text{I}.4.1)\left\{\begin{array}{ll}
C_0.B_{nm}\, x^{nm} = A_{nm}\, x^{nm} & \forall x \in X \\
C_1x.B_{nm}\, x^{nm} + C_0.B_{nm+1}\, x^{nm+1} = A_{nm+1}\, x^{nm+1} & \forall x \in X \\
\vdots & \\
C_nx^n.B_{nm}\, x^{nm} + \ldots + C_0.B_{nm+n}\, x^{nm+n} = A_{nm+n}\, x^{nm+n} & \forall x \in X
\end{array}\right.
$$

with $B_{nm+j}\, x^{nm+j} \equiv 0$ for $j > m$

$$
(\text{I}.4.2)\left\{\begin{array}{l}
C_{n+1}\, x^{n+1}.B_{nm}\, x^{nm} + \ldots + C_{n+1-m}\, x^{n+1-m}.B_{nm+m}\, x^{nm+m} = 0 \quad \forall x \in X \\
\vdots \\
C_{n+m}\, x^{n+m}.B_{nm}\, x^{nm} + \ldots + C_n\, x^n.B_{nm+m}\, x^{nm+m} = 0 \quad \forall x \in X
\end{array}\right.
$$

with $C_k\, x^k \equiv 0$ for $k < 0$.

A solution of (I.4.2) can be computed by means of the following determinants in Y; these formulas are direct generalizations of the classical formulas for the solution of a homogeneous system.

$$
B_{nm}\, x^{nm} = \begin{vmatrix}
C_n\, x^n & \ldots & C_{n+1-m}\, x^{n+1-m} \\
C_{n+1}\, x^{n+1} & \ldots & C_{n+2-m}\, x^{n+2-m} \\
\vdots & & \vdots \\
C_{n+m-1}\, x^{n+m-1} & \ldots & C_n\, x^n
\end{vmatrix} \in L(X^{nm}, Y)
$$

$$
B_{nm+j}\, x^{nm+j} = \begin{vmatrix}
C_n\, x^n & \ldots & \boxed{-C_{n+1}\, x^{n+1}} & \ldots & C_{n+1-m}\, x^{n+1-m} \\
\vdots & & \vdots & & \vdots \\
C_{n+m-1}\, x^{n+m-1} & \ldots & -C_{n+m}\, x^{n+m} & \ldots & C_n\, x^n
\end{vmatrix} \in L(X^{nm+j}, Y) \quad \text{for } 1 \le j \le m
$$

j^{th} column in $B_{nm}\, x^{nm}$ replaced by this column

For every solution of (I.4.2) a solution of (I.4.1) can be calculated by substitution of the $B_{nm+j}\, x^{nm+j}$ $(j=0,\ldots,m)$ in the left hand side of (I.4.1).
So, using the classical formulas, we get immediately the shift of degrees by n.m in P(x) and Q(x). A second argument for the choice of (P(x),Q(x)) will be given at the end of this paragraph. For the moment we want to give some more determinant representations. When we calculate a solution of (I.4.2) and (I.4.1) by means of the determinants above, we will denote it by $(P_{[n,m]}(x), Q_{[n,m]}(x))$. So

$$Q_{[n,m]}(x) = \begin{vmatrix} I & \cdots & & I \\ C_{n+1}x^{n+1} & C_n x^n & \cdots & C_{n+1-m}x^{n+1-m} \\ C_{n+2}x^{n+2} & \cdots & & \vdots \\ \vdots & & & \vdots \\ C_{n+m}x^{n+m} & \cdots & & C_n x^n \end{vmatrix} \qquad (I.4.3)$$

$$P_{[n,m]}(x) = \begin{vmatrix} F_n(x) & F_{n-1}(x) & \cdots & F_{n-m}(x) \\ C_{n+1}x^{n+1} & \cdots & & C_{n+1-m}x^{n+1-m} \\ \vdots & & & \vdots \\ C_{n+m}x^{n+m} & \cdots & & C_n x^n \end{vmatrix} \qquad (I.4.4)$$

$$(F.Q_{[n,m]} - P_{[n,m]})(x) = \begin{vmatrix} \overline{F}_{n+m}(x) & \overline{F}_{n+m-1}(x) & \cdots & \overline{F}_n(x) \\ C_{n+1}x^{n+1} & \cdots & & C_{n+1-m}x^{n+1-m} \\ \vdots & & & \vdots \\ C_{n+m}x^{n+m} & \cdots & & C_n x^n \end{vmatrix} \qquad (I.4.5)$$

where $F_i(x) = \sum_{k=0}^{i} C_k x^k$ and $F_i(x) \equiv 0$ for $i < 0$ and $\overline{F}_i(x) = F(x) - F_i(x)$.

These formulas are also direct generalizations of the classical formulas for univariate Padé approximants.

Remark also the fact that if we calculate the (n,o) abstract Padé approximant for F, we find the nth partial sum of the abstract Taylor series. For if $B_{nm} = I$ then $A_i x^i = C_i x^i$, $i=o,\ldots,n$ is a solution of system (I.4.1).

Let's again take a look at the nonlinear operator

$$F : \mathbb{R}^2 \to \mathbb{R}^2 : \begin{pmatrix} x \\ y \end{pmatrix} \to \begin{pmatrix} 1+x+\sin(xy) \\ x^2+y^2-4xy \end{pmatrix} = \begin{pmatrix} 1 \\ 0 \end{pmatrix} + \begin{pmatrix} x \\ 0 \end{pmatrix} + \begin{pmatrix} xy \\ x^2+y^2-4xy \end{pmatrix} + \ldots$$

Take $n = 1$ and $m = 2$. Using (I.4.3) and (I.4.4) we find

$$Q_{[1,2]}(x,y) = \begin{vmatrix} \binom{1}{1} & \binom{1}{1} & \binom{1}{1} \\ \begin{pmatrix} xy \\ x^2+y^2-4xy \end{pmatrix} & \binom{x}{0} & \binom{1}{0} \\ \binom{0}{0} & \begin{pmatrix} xy \\ x^2+y^2-4xy \end{pmatrix} & \binom{x}{0} \end{vmatrix} = \begin{pmatrix} x(x-y) + x^2y(y-1) \\ (x^2+y^2-4xy)^2 \end{pmatrix}$$

$$P_{[1,2]}(x,y) = \begin{vmatrix} \begin{pmatrix} 1+x \\ 0 \end{pmatrix} & \binom{1}{0} & \binom{0}{0} \\ \begin{pmatrix} xy \\ x^2+y^2-4xy \end{pmatrix} & \binom{x}{0} & \binom{1}{0} \\ \binom{0}{0} & \begin{pmatrix} xy \\ x^2+y^2-4xy \end{pmatrix} & \binom{x}{0} \end{vmatrix} = \begin{pmatrix} x^2(1+x-y) - xy(1+x) \\ 0 \end{pmatrix}$$

which is clearly a nontrivial solution of (I.4.1) and (I.4.2).
When we would try for n=1 and m=2 to find a couple of abstract polynomials $(P(x,y),Q(x,y)) = (A_1\binom{x}{y} + A_0\ ,\ B_2\binom{x}{y}^2 + B_1\binom{x}{y} + B_0)$ such that $(F.Q-P)(x,y) = O(\binom{x}{y}^{n+m+1}) = O(\binom{x}{y}^4)$, not working with the shift of degrees by n.m=2, we would remark that this problem has only the solution $Q(x,y) \equiv 0 \equiv P(x,y)$, which is not very useful. The reason is that we have now an overdetermined homogeneous system. More about multivariate Padé approximants can be found in chapter II.

§ 5. RELATIONS BETWEEN (P,Q) and ($P_\star,Q_\star$)

5.1. *Order and degree of P, Q, $P_\star$ and $Q_\star$*

From now on we will use the notations $P(x) = \sum_{i=0}^{n} A_{nm+i}\ x^{nm+i}$ and $Q(x) = \sum_{j=0}^{m} B_{nm+j}\ x^{nm+j}$ for solutions of (I.3.1), $P_\star(x) = \sum_{i=\partial_0 P_\star}^{\partial P_\star} A_{\star i}\ x^i$ and $Q_\star(x) = \sum_{j=\partial_0 Q_\star}^{\partial Q_\star} B_{\star j}\ x^j$ for the numerator and denominator of a reduced rational form of $\frac{1}{Q}.P$ and $T(x) = \sum_{k=t_0}^{\partial T} T_k\ x^k$ for the polynomial such that $P = P_\star.T$ and $Q = Q_\star.T$ where $t_0 = \partial_0 T$.
We will now give a few simple theorems about solutions (P,Q) of (I.3.1) and about the $(P_\star,Q_\star)$. Similar theorems exist for the univariate Padé approximants.

Theorem I.5.1.:

a) Let (P,Q) satisfy (I.3.1). Then $\partial_o P \geq \partial_o Q$.

b) Let $\frac{1}{Q_\star}.P_\star$ be a reduced form of $\frac{1}{Q}.P$. If $D(T_{t_o}) \neq \emptyset$ or $\partial_o Q_\star = o$ then $\partial_o P_\star \geq \partial_o Q_\star$.

Proof:

The proof of a) is very simple because of the equivalence of (I.3.1) with the systems (I.4.1) and (I.4.2).

The proof of b) is similar.

If $\partial_o Q_\star = o$ then b) is automatically satisfied.

Suppose $\partial_o P_\star < \partial_o Q_\star$ when $\partial_o Q_\star > o$.

Then $\partial_o P < \partial_o Q$ because of lemma I.2.2 which we can apply to the first nontrivial term in P since $D(T_{t_o}) \neq \emptyset$. This is a contradiction with a).

■

We introduce the notion of pseudo-degree for polynomials without tail like the ones considered in definition I.3.3 if $n > o$ and $m > o$ and like $P_\star$ and $Q_\star$ if $\partial_o Q_\star > o$ and $D(T_{t_o}) \neq \emptyset$.

Definition I.5.1.:

a) $\partial_1 P = \partial P - \partial_o Q$ is called the pseudo-degree of P and $\partial_1 Q = \partial Q - \partial_o Q$ the pseudo-degree of Q; theorem I.5.1 a) justifies the term $-\partial_o Q$.

b) $\partial_1 P_\star = \partial P_\star - \partial_o Q_\star$ is called the pseudo-degree of $P_\star$ and $\partial_1 Q_\star = \partial Q_\star - \partial_o Q_\star$ the pseudo-degree of $Q_\star$, if $D(T_{t_o}) \neq \emptyset$ or $\partial_o Q_\star = o$; theorem I.5.1 b) justifies the term $-\partial_o Q_\star$.

When $\partial_o Q = o$ or $\partial_o Q_\star = o$ the pseudo-degrees of P and Q or $P_\star$ and $Q_\star$ equal the exact degrees.

Theorem I.5.2.:

a) $\partial_1 P \leq n$ and $\partial_1 Q \leq m$.

b) Let Y be a commutative Banach algebra without nilpotent elements.

If $D(T_{t_o}) \neq \emptyset$ or $Q_\star(0)$ is regular in Y, then $\partial_1 P_\star \leq n$ and $\partial_1 Q_\star \leq m$.

Proof:

a) $\partial_1 P = \partial P - \partial_0 Q \le nm+n - \partial_0 Q \le n$ since $\partial_0 Q \ge nm$.

$\partial_1 Q = \partial Q - \partial_0 Q \le nm+m - \partial_0 Q \le m$ since $\partial_0 Q \ge nm$.

b) $\partial_1 Q_\star = \partial Q_\star - \partial_0 Q_\star \le (\partial Q - \partial_0 T) - \partial_0 Q_\star$ since $D(T) \ne \emptyset$

$\le \partial Q - nm$ since $\partial_0 T + \partial_0 Q_\star = \partial_0 Q \ge nm$

$\le m$

$\partial_1 P_\star = \partial P_\star - \partial_0 Q_\star \le (\partial P - \partial_0 T) - \partial_0 Q_\star$ since $D(T) \ne \emptyset$

$\le \partial P - nm$ since $\partial_0 T + \partial_0 Q_\star = \partial_0 Q \ge nm$

$\le n$

If $Q_\star(0)$ is regular, the proof is similar because now $\partial_0 Q_\star = o$ and $\partial_0 T = \partial_0 Q$. ■

The following example proves the need of $D(T_{t_o}) \ne 0$ to conclude that $\partial_1 P_\star \le n$ and $\partial_1 Q_\star \le m$.

Take $F : \mathbb{R}^2 \to \mathbb{R}^2 : \binom{x}{y} \to \begin{pmatrix} \cos(\alpha - x + y) \\ \frac{xe^x - ye^y}{x-y} \end{pmatrix}$ with $\alpha \ne k\pi$ and take $n = 1 = m$.

The couple of abstract polynomials $(P_\star . T, Q_\star . T)$ with

$$P_\star\binom{x}{y} = \begin{pmatrix} \cos\alpha + (x-y)(\sin\alpha + 0.5\cot g\alpha\cos\alpha) \\ x+y+0.5(x^2+3xy+y^2) \end{pmatrix},$$

$$Q_\star\binom{x}{y} = \begin{pmatrix} 1+0.5(x-y)\cot g\alpha \\ x+y-0.5(x^2+xy+y^2) \end{pmatrix}$$

and $T\binom{x}{y} = \binom{0}{1} + \binom{y-x}{0}$, satisfies (I.3.1). Here $\partial_0 Q_\star = o$, $\partial P_\star = 2$, $\partial Q_\star = 2$ and $T_{t_o}\binom{x}{y} = \binom{0}{1}$. So $\partial_1 P_\star = 2 > 1 < 2 = \partial_1 Q_\star$.

5.2. *Order of $F.Q_\star - P_\star$*

The following theorem is frequently used in the proofs of further properties.

Theorem I.5.3.:

a) If $D(T_{t_0}) \neq \emptyset$, then $t_0 = nm - \partial_0 Q_\star + r$ with $r \geq 0$ and $(P_\star . T_{t_0}, Q_\star . T_{t_0})$ satisfies (I.3.1)

b) If Y contains no nilpotent elements then also $0 \leq r \leq \min(n-\partial_1 P_\star, m-\partial_1 Q_\star)$.

Proof:

a) Because $D(T_{t_0}) \neq \emptyset$, $t_0 + \partial_0 Q_\star = \partial_0 Q \geq nm$.

We write $t_0 = nm - \partial_0 Q_\star + r$ with $r \geq 0$.

Now $F(x).Q(x) - P(x) = T(x).[F(x).Q_\star(x) - P_\star(x)] = O(x^{nm+n+m+1})$.

If $T(x) = T_{t_0} x^{t_0} + \ldots$ with $D(T_{t_0}) \neq \emptyset$ then also

$T_{t_0} x^{t_0}.[F(x).Q_\star(x) - P_\star(x)] = O(x^{nm+n+m+1})$ because of the

equivalence of (I.3.1) with (I.4.1) and (I.4.2).

b) Because $D(T) \neq \emptyset$, we have according to lemma I.2.3:

$$\begin{cases} \partial P_\star \leq \partial(P_\star . T) - \partial_0 T \leq nm+n - (nm - \partial_0 Q_\star + r) \\ \partial Q_\star \leq \partial(Q_\star . T) - \partial_0 T \leq nm+m - (nm - \partial_0 Q_\star + r) \end{cases}$$

and so: $$\begin{cases} \partial_1 P_\star = \partial P_\star - \partial_0 Q_\star \leq n-r \\ \partial_1 Q_\star = \partial Q_\star - \partial_0 Q_\star \leq m-r \end{cases}$$

■

When we compare this theorem with the similar one for the classical univariate Padé approximant, we remark that the term nm in t_0 is due to the choice of the order of the couple of polynomials (P,Q) in definition I.3.3 and that the term $-\partial_0 Q_\star$ is due to the fact that not always $\partial_0 Q_\star = 0$.
We give some illustrative examples.

Let $F : \mathbb{R}^2 \to \mathbb{R}^2 : \binom{x}{y} \to \begin{pmatrix} \frac{1}{1-x} \\ e^{x+y} \end{pmatrix}$ and take n = 1, m = 2.

The couple of abstract polynomials $(P_\star . T, Q_\star . T)$ with

$P_\star \binom{x}{y} = \begin{pmatrix} 1 \\ 1 + \frac{1}{3}(x+y) \end{pmatrix}$, $Q_\star \binom{x}{y} = \begin{pmatrix} 1 - x \\ 1 - \frac{2}{3}(x+y) + \frac{1}{6}(x+y)^2 \end{pmatrix}$ and

$$T\binom{x}{y} = \begin{pmatrix} (1+x)\, L_2 \binom{x}{y}^2 + L_3 \binom{x}{y}^3 \\ \frac{(x+y)^2}{2} \end{pmatrix}$$ where $L_2 \in L(X^2,\mathbb{R})$, $L_3 \in L(X^3,\mathbb{R})$,

satisfies (I.3.1). So if $D(L_2) \neq \emptyset$ theorem I.5.3 is satisfied with $t_o = 2$,

$$T_2 \binom{x}{y}^2 = \begin{pmatrix} L_2 \binom{x}{y}^2 \\ (x+y)^2/2 \end{pmatrix}$$ and $r = o$.

Let $F : \mathbb{R}^2 \to \mathbb{R}^2 : \binom{x}{y} \to \begin{pmatrix} 1 + \sin(x+xy) \\ 1 + \frac{x}{0.1-y} + \sin(xy) \end{pmatrix}$ and take $n = 1$, $m = 2$.

The couple of abstract polynomials $(P_\star.T, Q_\star.T)$ with

$$P_\star \binom{x}{y} = \begin{pmatrix} x - y + \frac{5}{6}x^2 - 2xy \\ x - 1.01\, y + 10\, y^2 + 10\, x^2 - 20.2\, xy \end{pmatrix},$$

$$Q_\star \binom{x}{y} = \begin{pmatrix} x - y - xy - \frac{x^2}{6} + xy^2 + \frac{x^3}{6} \\ x - 1.01\, y + 10\, y^2 - 10.1\, xy + 2.01\, xy^2 \end{pmatrix}$$ and

$T\binom{x}{y} = \binom{x}{100\, x}$ satisfies (I.3.1).

So theorem I.5.3. is satisfied with $t_o = 1$ and $r = o$.

Let $F : \mathbb{R}^2 \to \mathbb{R}^2 : \binom{x}{y} \to \begin{pmatrix} \frac{1+y}{1+y^2} \\ 1 - \cos x \end{pmatrix}$ and take $n = 2$, $m = 1$.

The couple of abstract polynomials $(P_\star.T, Q_\star.T)$ with

$P_\star \binom{x}{y} = \begin{pmatrix} 1 + y \\ x^2/2 \end{pmatrix}$, $Q_\star \binom{x}{y} = \binom{1}{1}$ and $T\binom{x}{y} = \begin{pmatrix} 0 \\ -x^2/2 \end{pmatrix} + \begin{pmatrix} -y^3 \\ 0 \end{pmatrix}$

satisfies (I.3.1), but with $t_o = 2$, $D(T_{t_o}) = \emptyset$. It is easy to see that

$(P_\star.T_{t_o}, Q_\star.T_{t_o})$ does not satisfy (I.3.1).

Let $F : C'([1,T]) \to C([1,T]) : x(t) \to e^{x(t)} \frac{dx}{dt} - (1+c)$ with c a small nonnegative number. Take $n = 1 = m$. The couple of abstract polynomials $(P_\star.T, Q_\star.T)$ with $P_\star(x) = \frac{dx}{dt} - (1+c)(1-x(t))$, $Q_\star(x) = 1-x(t)$ and $T(x) = \frac{dx}{dt}$ satisfies (I.3.1). Theorem 1.5.3 is satisfied with $t_o = 1$ and $r = o$.

Theorem I.5.4.:

a) If $D(T_{t_o}) \neq \emptyset$ then $(F.Q_\star - P_\star)(x) = O(x^{nm+n+m+1-t_o})$

b) If Y contains no nilpotent elements and $D(T_{t_o}) \neq \emptyset$ then

$(F.Q_\star - P_\star)(x) = O(x^{\partial_o Q_\star + \partial_1 P_\star + \partial_1 Q_\star + 1})$

Proof:

a) Suppose $\partial_o(F.Q_\star - P_\star) = j$ with $j < nm+n+m+1-t_o$.

Now since $D(T_{t_o}) \neq \emptyset$: $(P_\star . T_{t_o}, Q_\star . T_{t_o})$ satisfies (I.3.1).

So $[(F.Q_\star - P_\star).T_{t_o}](x) = O(x^{nm+n+m+1})$ and consequently

$nm+n+m+1 \leq j + t_o$ which is a contradiction.

b) Suppose $\partial_o(F.Q_\star - P_\star) = j$ with $j < \partial_o Q_\star + \partial_1 P_\star + \partial_1 Q_\star + 1$.

Then for every integer r with $o \leq r \leq \min(n - \partial_1 P_\star, m - \partial_1 Q_\star)$,

for $t_o = nm - \partial_o Q_\star + r$:

$$\partial_o Q_\star + \partial_1 P_\star + \partial_1 Q_\star + 1 + (nm - \partial_o Q_\star + r) > j + nm - \partial_o Q_\star + r$$
$$\geq nm+n+m+1,$$

which is in contradiction with theorem I.5.3 b) since

$\partial_1 P_\star \leq n$ and $\partial_1 Q_\star \leq m$ and

$\partial_o Q_\star + \partial_1 P_\star + \partial_1 Q_\star + 1 + (nm - \partial_o Q_\star + r) \leq nm+n+m+1$ ∎

It is also easy to see that if $D(T_{t_o}) \neq \emptyset$, then every couple of abstract polynomials $(P_\star . L, Q_\star . L)$, with L a bounded t_o-linear operator such that $D(L) \cap (D(P_\star) \cup D(Q_\star)) \neq \emptyset$, satisfies (I.3.1).

The fact that $(F.Q_\star - P_\star)(x) = O(x^j)$ with j given by theorem I.5.4 implies that

$(F.Q_\star - P_\star)^{(i)}(0) \equiv 0$ for $i = o, \ldots, j-1$ at least

For polynomials $P_\star$ and $Q_\star$ with $\partial_o P_\star \geq \partial_o Q_\star$ we know that always

$(F.Q_\star - P_\star)^{(i)}(0) \equiv 0$ for $i = o, \ldots, \partial_o Q_\star - 1$

So the meaningful relations are

$(F.Q_\star - P_\star)^{(i)}(0) \equiv 0$ for $i = \partial_o Q_\star, \ldots, j-1$ at least (I.5.1)

When $0 \in D(Q_\star)$ and thus $\partial_o Q_\star = o$, the relations (I.5.1) can be rewritten as

$$F^{(i)}(0) = \left(\frac{1}{Q_\star}.P_\star\right)^{(i)}(0) \text{ for } i = o, \ldots, j-1 \text{ at least.}$$

So (I.5.1) clearly has an interpolatory meaning in 0.

§ 6. COVARIANCE PROPERTIES

Since the (n,m) abstract Padé approximant is an equivalence-class containing couples of abstract polynomials, we are going to represent it by one of its elements; for the sake of simplicity we will denote this representant also by $(P_\star, Q_\star)$.
Let the operator $P_{n,m}$ (for n and m chosen) associate with the operator F the equivalence-class of the $(P_\star, Q_\star)$. We are looking for operators Φ working on F that commute more or less with the Padé operator $P_{n,m}$:

$$\Phi[P_{n,m}(F)] = P_{n_\Phi, m_\Phi}[\Phi(F)]$$

with n_Φ and m_Φ depending on the considered Φ.
The first property we are going to prove is the reciprocal covariance of abstract Padé approximants.

Theorem I.6.1.:

> Suppose $0 \in D(F)$. If $(P_\star, Q_\star)$ is the (n,m) abstract Padé approximant for F, then $(Q_\star, P_\star)$ is the (m,n) abstract Padé approximant for $\frac{1}{F}$.

Proof:

Since $0 \in D(F)$, an open ball $B(0,r)$ exists where $\frac{1}{F}$ is defined.
If $(P_\star, Q_\star)$ is the (n,m) abstract Padé-approximant for F, an abstract polynomial T exists such that
$(P,Q) = (P_\star.T, Q_\star.T)$ satisfies (I.3.1) for F and $D(P) \cup D(Q) \neq \emptyset$.
In other words $(F.Q-P)(x) = O(x^{nm+n+m+1})$
This implies that $(\frac{1}{F}.P-Q)(x) = O(x^{nm+n+m+1})$ since $\frac{1}{F}$ is abstract analytic in a neighbourhood of 0.
So $(Q,P) = (Q_\star.T, P_\star.T)$ satisfies (I.3.1) for $\frac{1}{F}$ and $D(Q) \cup D(P) \neq \emptyset$.

■

Theorem I.6.2.:

Suppose a, b, c, d $\in$ Y and 0 $\in$ D(c.F+d). If $(P_\star,Q_\star)$ is the (n,n) abstract Padé approximant for F and D(c.P+d.Q) $\cup$ D(a.P+b.Q) $\neq \emptyset$, then the (n,n) abstract Padé approximant for $\frac{1}{c.F+d}.(a.F+b)$ is $(a.P_\star+b.Q_\star, c.P_\star+d.Q_\star)$.

Proof:

Since 0 $\in$ D(c.F+d), an open ball B(0,r) exists where $\frac{1}{c.F+d}.(a.F+b)$ is defined.

If $(P_\star,Q_\star)$ is the (n,n) abstract Padé-approximant for F, an abstract polynomial T exists such that $(P,Q) = (P_\star.T,\ Q_\star.T)$ satisfies (I.3.1) for F and D(P) $\cup$ D(Q) $\neq \emptyset$.

In other words $(F.Q-P)(x) = O(x^{n^2+2n+1})$.

Now $\partial_0(a.P+b.Q) \geq n^2$ since $\partial_0 P \geq n^2$ and $\partial_0 Q \geq n^2$ and $\partial(a.P+b.Q) \leq n^2 + n$ since $\partial P \leq n^2 + n$ and $\partial Q \leq n^2 + n$.

Also $\partial_0(c.P+d.Q) \geq n^2$ and $\partial(c.P+d.Q) \leq n^2 + n$.

Since $(F.Q-P)(x) = O(x^{n^2+2n+1})$ and 0 $\in$ D(c.F+d), also

$$[(a.d-b.c).\frac{1}{c.F+d}.(F.Q-P)](x) = O(x^{n^2+2n+1}).$$

Now $\frac{1}{c.F+d}.(a.F+b).(c.P+d.Q) - (a.P+b.Q) = \frac{1}{c.F+d}.(F.Q-P)(a.d-b.c)$

and consequently

$$\frac{1}{c.F+d}.(a.F+b).(c.P+d.Q) - (a.P+b.Q) = O(x^{n^2+2n+1}).$$

We already know that D(c.P+d.Q) $\cup$ D(a.P+b.Q) $\neq \emptyset$.

■

We remark that if $(P_\star,Q_\star)$ were the (n,m) abstract Padé approximant for F with n > m for instance, then a.P+b.Q was indeed an abstract polynomial of order at least nm and degree at most nm+n but c.P+d.Q not necessarily an abstract polynomial of degree at most nm+m. This explains the condition in theorem I.6.2 that $(P_\star,Q_\star)$ is the (n,n) abstract Padé approximant for F.

Theorem I.6.3.:

Suppose $A \in L(X,X)$ and A^{-1} exists. If $(P_\star,Q_\star)$ is the (n,m) abstract Padé approximant for F and if $R_\star(x):=P_\star(Ax)$, $S_\star(x):=Q_\star(Ax)$, $G(x):=F(Ax)$, then $(R_\star,S_\star)$ is the (n,m) abstract Padé approximant for G.

Proof:

If $L \in L(X^i,Y)$, then $L \circ A \in L(X^i,Y)$ when defined by $(L \circ A)x^i = L(Ax)^i$ [6 pp. 289].

Because $(P_\star,Q_\star)$ is the (n,m) abstract Padé-approximant for F, an abstract polynomial T exists such that $(P,Q) = (P_\star.T,\ Q_\star.T)$ satisfies (I.3.1) for F and $D(P) \cup D(Q) \neq \emptyset$.

In other words, there exist nonnegative constants $r < 1$ and K such that $\|(F.Q-P)(x)\| \leq K.\|x\|^{nm+n+m+1}$ for $\|x\| < r$.

Let $S(x) = T(Ax).S_\star(x)$ and $R(x) = T(Ax).R_\star(x)$.

Then $\|(G.S-R)(x)\| = \|(F.Q-P)(Ax)\| \leq K.\|Ax\|^{nm+n+m+1}$

$\leq (K.\|A\|^{nm+n+m+1})\ \|x\|^{nm+n+m+1}$

for $\|x\| < r$.

Thus $(G.S-R)(x) = O(x^{nm+n+m+1})$.

Since $D(R) = \{x \in X \mid R(x) \text{ is regular in } Y\}$

$= \{x \in X \mid P(Ax) \text{ is regular in } Y\}$

$= \{A^{-1}x \mid x \in D(P)\}$

$= A^{-1}(D(P))$

and $D(S) = A^{-1}(D(Q))$

we can conclude that $D(R) \cup D(S) = A^{-1}[D(P) \cup D(Q)] \neq \emptyset$.

∎

This theorem has two important consequences: the scale covariance of abstract Padé approximants formulated in corollary I.6.1 and the conservation of symmetry formulated in corollary I.6.2.

Corollary I.6.1.:

Let $\lambda \in \Lambda \setminus \{0\}$. If $(P_\star,Q_\star)$ is the (n,m) abstract Padé approximant for F and $R_\star(x):=P_\star(\lambda x)$, $S_\star(x):=Q_\star(\lambda x)$, $G(x):=F(\lambda x)$, then $(R_\star,S_\star)$ is the (n,m) abstract Padé approximant for G.

Corollary I.6.2.:

Let $X = X_1 \times X_2$ and $F(x_1,x_2) = F(x_2,x_1)$. If $(P_\star,Q_\star)$ is the (n,m) abstract Padé approximant for F, then $(P_\star(x_1,x_2),Q_\star(x_1,x_2)) \sim (P_\star(x_2,x_1),Q_\star(x_2,x_1))$.

§ 7. RECURRENCE RELATIONS

7.1. *Two-term identities*

Frobenius [20] supplied most of the identities for the classical Padé approximants. We will now discuss their generalizations. The first group of identities we will consider are the two-term identities. By definition I.3.3 we can write

$$(F.Q_{[n,m]} - P_{[n,m]})(x) = O(x^{nm+n+m+1})$$

$$(F.Q_{[n+1,m]} - P_{[n+1,m]})(x) = O(x^{(n+1)m+n+m+2})$$

$$(F.Q_{[n,m+1]} - P_{[n,m+1]})(x) = O(x^{n(m+1)+n+m+2})$$

$$(F.Q_{[n+1,m+1]} - P_{[n+1,m+1]})(x) = O(x^{(n+1)(m+1)+n+m+3})$$

Thus

$$(P_{[n+1,m]}\, Q_{[n,m]} - P_{[n,m]}\, Q_{[n+1,m]})(x) =$$

$$= \left[(F.Q_{[n,m]} - P_{[n,m]}).Q_{[n+1,m]} - (F.Q_{[n+1,m]} - P_{[n+1,m]}).Q_{[n,m]}\right](x)$$

$$= O(x^{nm+(n+1)m+n+m+1})$$

While

$$\partial(P_{[n+1,m]}\, Q_{[n,m]} - P_{[n,m]}\, Q_{[n+1,m]}) \le nm+(n+1)m+n+m+1$$

and analogously

$$(P_{[n,m+1]}\, Q_{[n,m]} - P_{[n,m]}\, Q_{[n,m+1]})(x) = O(x^{nm+n(m+1)+n+m+1})$$

$$\partial(P_{[n,m+1]}\, Q_{[n,m]} - P_{[n,m]}\, Q_{[n,m+1]}) \le nm+n(m+1)+n+m+1$$

$$(P_{[n,m+1]}\, Q_{[n+1,m]} - P_{[n+1,m]}\, Q_{[n,m+1]})(x) = O(x^{(n+1)m+n(m+1)+n+m+2})$$

$$\partial(P_{[n,m+1]}\, Q_{[n+1,m]} - P_{[n+1,m]}\, Q_{[n,m+1]}) \le n(m+1)+(n+1)m+n+m+2$$

$$(P_{[n+1,m+1]}\, Q_{[n,m]} - P_{[n,m]}\, Q_{[n+1,m+1]})(x) = O(x^{nm+(n+1)(m+1)+n+m+1})$$

$$\partial(P_{[n+1,m+1]}\, Q_{[n,m]} - P_{[n,m]}\, Q_{[n+1,m+1]}) \le nm+(n+1)(m+1)+n+m+1$$

Let us introduce the notation

$$H_j\,(S_i) = \begin{vmatrix} S_i & \cdots & S_{i-j+1} \\ \vdots & \ddots & \vdots \\ S_{i+j-1} & \cdots & S_i \end{vmatrix}$$

for these determinants where the S_i $(i=0,1,\ldots)$ are elements in the commutative Banach algebra Y. Then because of the formulas (I.4.3), (I.4.4) and (I.4.5) we have

$$(P_{[n+1,m]}\, Q_{[n,m]} - P_{[n,m]}\, Q_{[n+1,m]})(x) = (-1)^m\, H_{m+1}(C_{n+1}x^{n+1}).H_m(C_{n+1}x^{n+1}) \quad (I.7.1)$$

$$(P_{[n,m+1]}\, Q_{[n,m]} - P_{[n,m]}\, Q_{[n,m+1]})(x) = (-1)^m\, H_{m+1}(C_{n+1}x^{n+1}).H_{m+1}(C_n x^n) \quad (I.7.2)$$

$$(P_{[n,m+1]}\, Q_{[n+1,m]} - P_{[n+1,m]}\, Q_{[n,m+1]})(x) = (-1)^m\, [H_{m+1}(C_{n+1}x^{n+1})]^2 \quad (I.7.3)$$

$$(P_{[n+1,m+1]}\, Q_{[n,m]} - P_{[n,m]}\, Q_{[n+1,m+1]})(x) = (-1)^m\, [H_{m+1}\,(C_{n+1}x^{n+1})]^2 \quad (I.7.4)$$

Fo (I.7.3) we have used Silvester's identity [2, pp.15] which is also valid in Y and states that

$$H_{m+1}\,(C_{n+2}x^{n+2}).H_{m+1}\,(C_n x^n)+H_{m+2}(C_{n+1}x^{n+1}).H_m\,(C_{n+1}x^{n+1}) =$$

$$= [H_{m+1}\,(C_{n+1}\,x^{n+1})]^2$$

These two-term identities will be used to prove other recursion relations in § 5. of chapter II. The second group of identities we will consider are the five-term identities.

7.2. *The ε-algorithm*

The ε-algorithm of Wynn is closely related to the Padé approximants of a univariate function in the following sense: if we apply the ε-algorithm to the partial sums of the power series

$$F(x) = \sum_{k=0}^{\infty} c_k x^k$$

then $\varepsilon_{2m}^{(n-m)}$ is the (n,m) Padé approximant for F(x) where n is the degree of the numerator and m is the degree of the denominator [9 pp. 66-68].
We shall now see that the (n,m) abstract Padé approximant satisfies the same property, but first of all we briefly repeat the nonlinear ε-algorithm. Input are the elements of a sequence $\{S_i \mid i=0,1,\dots\}$; let us take the sequence in Y.
The following computations are performed:

$$\varepsilon_{-1}^{(i)} = 0 \qquad i = 0,\ 1,\ \dots$$

$$\varepsilon_{o}^{(i)} = S_i \qquad i = 0,\ 1,\ \dots$$

$$\varepsilon_{2j}^{(-j-1)} = 0 \quad j = 0,\ 1,\ \dots$$

$$\varepsilon_{j+1}^{(i)} = \varepsilon_{j-1}^{(i+1)} + [\varepsilon_j^{(i+1)} - \varepsilon_j^{(i)}]^{-1} \qquad \begin{array}{l} j = 0,\ 1,\ \dots \\ i = -j,\ -j+1,\ \dots \end{array}$$

The $\varepsilon_j^{(i)}$ can be ordered in a table where (i) indicates a diagonal and j a column:

$$\begin{array}{lllll}
 & \varepsilon_o^{(-1)} = 0 & & \varepsilon_2^{(-2)} = 0 & \dots \\
\varepsilon_{-1}^{(o)} = 0 & & \varepsilon_1^{(-1)} & & \\
 & \varepsilon_o^{(o)} = S_o & & \varepsilon_2^{(-1)} & \dots \\
\varepsilon_{-1}^{(1)} = 0 & & \varepsilon_1^{(o)} & & \\
 & \varepsilon_o^{(1)} = S_1 & & \varepsilon_2^{(o)} & \dots \\
\varepsilon_{-1}^{(2)} = 0 & & \varepsilon_1^{(1)} & & \\
 & \varepsilon_o^{(2)} = S_2 & & \varepsilon_2^{(1)} & \dots \\
\varepsilon_{-1}^{(3)} = 0 & \vdots & \varepsilon_1^{(2)} & \vdots & \\
\vdots & & \vdots & &
\end{array}$$

We introduce the notations

$$\Delta S_i = S_{i+1} - S_i$$

and

$$\Delta^2 S_i = \Delta S_{i+1} - \Delta S_i$$

to prove the following property for the $\varepsilon_j^{(i)}$.
The proof is very technical and similar to the proof in [8 pp. 44-46] for $X = \mathbb{R} = Y$.

Lemma I.7.1.:

If $H_{j-1}(\Delta^2 S_{i+j-1})$ and $H_j(\Delta^2 S_{i+j-1})$ are regular in Y, then

$$\varepsilon_{2j}^{(i)} = \frac{\begin{vmatrix} S_{i+j} & \cdots & & S_i \\ \Delta S_{i+j} & \cdots & \Delta S_{i+1} & \Delta S_i \\ \vdots & \ddots & \vdots & \vdots \\ \Delta S_{i+2j-1} & \cdots & \Delta S_{i+j} & \Delta S_{i+j-1} \end{vmatrix}}{\begin{vmatrix} I & \cdots & I \\ \Delta S_{i+j} & \cdots & \Delta S_i \\ \vdots & & \vdots \\ \Delta S_{i+2j-1} & \cdots & \Delta S_{i+j-1} \end{vmatrix}}$$

and if $H_j(\Delta S_{i+j})$ and $H_{j+1}(\Delta S_{i+j})$ are regular in Y, then

$$\varepsilon_{2j+1}^{(i)} = \frac{\begin{vmatrix} I & \cdots & I \\ \Delta^2 S_{i+j} & \cdots & \Delta^2 S_i \\ \vdots & & \vdots \\ \Delta^2 S_{i+2j-1} & \cdots & \Delta^2 S_{i+j-1} \end{vmatrix}}{\begin{vmatrix} \Delta S_{i+j} & \cdots & \Delta S_i \\ \Delta^2 S_{i+j} & \cdots & \Delta^2 S_i \\ \vdots & & \vdots \\ \Delta^2 S_{i+2j-1} & \cdots & \Delta^2 S_{i+j-1} \end{vmatrix}}$$

with $S_i = 0$ for $i < 0$.

Of course we restrict ourselves to the case that the $\varepsilon_j^{(i)}$ are finite; since the ε-algorithm is a nonlinear algorithm, it can always happen that $\varepsilon_{j+1}^{(i)}$ does not exist (when $\varepsilon_j^{(i+1)} - \varepsilon_j^{(i)}$ is not regular in Y).

It is easy to see now that for $S_i = F_i(x) = \sum_{k=0}^{i} C_k x^k$ we get the following theorem.

Theorem I.7.1.:

If $D[H_{m-1}(\Delta^2 F_{n-1})] \neq \emptyset$ and $D[H_m(\Delta^2 F_{n-1})] \neq \emptyset$,

then

$$\varepsilon_{2m}^{(n-m)} = \frac{\begin{vmatrix} F_n(x) & \cdots & F_{n-m}(x) \\ C_{n+1}x^{n+1} & \cdots & C_{n-m+1}x^{n-m+1} \\ \vdots & & \vdots \\ C_{n+m}x^{n+m} & \cdots & C_n x^n \end{vmatrix}}{\begin{vmatrix} I & \cdots & I \\ C_{n+1}x^{n+1} & \cdots & C_{n-m+1}x^{n-m+1} \\ \vdots & & \vdots \\ C_{n+m}x^{n+m} & \cdots & C_n x^n \end{vmatrix}}$$

Numerator and denominator of $\varepsilon_{2m}^{(n-m)}$ are the determinantal formulas (I.4.4) and (I.4.3) for P(x) and Q(x), a solution of the Padé approximation problem of order (n,m).
Let us illustrate this by calculating part of the ε-table for the following nonlinear operator

$$F: C'([1,T]) \to C([1,T]): x(t) \to e^{x(t)} \frac{dx}{dt} - (1+c)$$

with c a small nonnegative constant.
The Taylor series expansion is

$$F(x) = \frac{dx}{dt} \sum_{k=0}^{\infty} \frac{1}{k!}[x(t)]^k - (1+c)$$

For the ε-table we get

$$
\begin{array}{cccc}
 & 0 & & 0 \\
0 & & \frac{-1}{1+c} & \\
 & -(1+c) & & \frac{-(1+c)^2}{1+c+\frac{dx}{dt}} \\
0 & & \frac{1}{\frac{dx}{dt}} & \\
 & \frac{dx}{dt}-(1+c) & & \frac{\frac{dx}{dt}-(1+c)(1-x(t))}{1-x(t)} \\
0 & & \frac{1}{x(t)\frac{dx}{dt}} & \\
 & \frac{dx}{dt}(1+x(t))-(1+c) & & \frac{\frac{dx}{dt}(1+\frac{1}{2}x(t))-(1+c)(1-\frac{1}{2}x(t))}{1-\frac{1}{2}x(t)} \\
0 & & \frac{1}{\frac{1}{2}x^2(t)\frac{dx}{dt}} & \vdots \\
\vdots & \frac{dx}{dt}(1+x(t)+\frac{1}{2}x^2(t))-(1+c) & \vdots & \\
 & \vdots & &
\end{array}
$$

It is easy to see that the odd columns are only intermediate results. By eliminating the odd columns, the ε-algorithm for the even columns can be rewritten as follows:

$$\varepsilon_{-2}^{(i)} = \infty \qquad (\text{i.e. } [\varepsilon_{-2}^{(i)}]^{-1} = 0) \qquad i = o, 1, \ldots$$

$$\varepsilon_{0}^{(i)} = S_i \qquad i = o, 1, \ldots$$

$$\varepsilon_{2j}^{(-j-1)} = 0 \qquad j = o, 1, \ldots$$

$$[\varepsilon_{2j+2}^{(i-1)} - \varepsilon_{2j}^{(i)}] = [\varepsilon_{2j+1}^{(i)} - \varepsilon_{2j+1}^{(i-1)}]^{-1}$$

$$= [\varepsilon_{2j-1}^{(i+1)} + (\varepsilon_{2j}^{(i+1)} - \varepsilon_{2j}^{(i)})^{-1} - \varepsilon_{2j-1}^{(i)} - (\varepsilon_{2j}^{(i)} - \varepsilon_{2j}^{(i-1)})^{-1}]^{-1}$$

$$= [(\varepsilon_{2j}^{(i+1)} - \varepsilon_{2j}^{(i)})^{-1} + (\varepsilon_{2j}^{(i-1)} - \varepsilon_{2j}^{(i)})^{-1} - (\varepsilon_{2j-2}^{(i+1)} - \varepsilon_{2j}^{(i)})^{-1}]^{-1}$$

and thus

$$[\varepsilon_{2j+2}^{(i-1)} - \varepsilon_{2j}^{(i)}]^{-1} + [\varepsilon_{2j-2}^{(i+1)} - \varepsilon_{2j}^{(i)}]^{-1} = [\varepsilon_{2j}^{(i+1)} - \varepsilon_{2j}^{(i)}]^{-1} + [\varepsilon_{2j}^{(i-1)} - \varepsilon_{2j}^{(i)}]^{-1} \qquad \begin{array}{l} j=o,1,\ldots \\ i=-j,-j+1,\ldots \end{array}$$

So we have the following relation between abstract Padé approximants.

Theorem I.7.2.:

If $H_{j-2}(\Delta^2 s_{i+j-1})$, $H_{j-1}(\Delta^2 S_{i+j-2})$, $H_{j-1}(\Delta^2 S_{i+j-1})$, $H_{j-1}(\Delta^2 S_{i+j})$, $H_j(\Delta^2 S_{i+j-2})$, $H_j(\Delta^2 S_{i+j-1})$, $H_j(\Delta^2 S_{i+j})$ and $H_{j+1}(\Delta^2 S_{i+j-1})$ are regular in Y then

$$\left[\frac{P_{[i+j,j+1]}}{Q_{[i+j,j+1]}} - \frac{P_{[i+j,j]}}{Q_{[i+j,j]}}\right]^{-1} + \left[\frac{P_{[i+j,j-1]}}{Q_{[i+j,j-1]}} - \frac{P_{[i+j,j]}}{Q_{[i+j,j]}}\right]^{-1} =$$

$$\left[\frac{P_{[i+j+1,j]}}{Q_{[i+j+1,j]}} - \frac{P_{[i+j,j]}}{Q_{[i+j,j]}}\right]^{-1} + \left[\frac{P_{[i+j-1,j]}}{Q_{[i+j-1,j]}} - \frac{P_{[i+j,j]}}{Q_{[i+j,j]}}\right]^{-1}$$

We are going to illustrate this by means of the ε-table we just calculated for the nonlinear operator F. Take j=1 and i=o and calculate

$$\varepsilon_4^{(-1)} = \frac{-(\frac{dx}{dt})^2+\frac{dx}{dt}(1+c)(-2x(t)+1)+(1+c)^2x(t)(1-\frac{1}{2}x(t))}{\frac{dx}{dt}(-1+x(t)-\frac{1}{2}x^2(t))-(1+c)x(t)(1-\frac{1}{2}x(t))}$$

We have then the following cross of abstract Padé approximants

$$\frac{-(1+c)^2}{1+c+\frac{dx}{dt}}$$

$$\frac{dx}{dt}-(1+c) \qquad \frac{\frac{dx}{dt}-(1+c)(1-x(t))}{1-x(t)} \qquad \frac{-(\frac{dx}{dt})^2+\frac{dx}{dt}(1+c)(-2x(t)+1)+(1+c)^2x(t)(1-\frac{1}{2}x(t))}{\frac{dx}{dt}(-1+x(t)-\frac{1}{2}x^2(t))-(1+c)x(t)(1-\frac{1}{2}x(t))}$$

$$\frac{\frac{dx}{dt}(1+\frac{1}{2}x(t))-(1+c)(1-\frac{1}{2}x(t))}{1-\frac{1}{2}x(t)}$$

In the chapters II and III the ε-algorithm is frequently used in numerical calculations.

7.3. *The qd-algorithm*

It is well-known that the quotient-difference algorithm can be used to construct univariate Padé approximants. We will first repeat it in an equivalent but slightly different way than usual. Only, this approach can be generalized when F is a nonlinear operator from a Banach space X into a commutative Banach algebra Y.
Let us consider a nonlinear real-valued function F of one real variable analytic in the origin:

$$F(x) = \sum_{k=o}^{\infty} c_k x^k \text{ with } c_k = \frac{1}{k!} F^{(k)}(0)$$

Let the series F be normal:

$$\begin{vmatrix} c_\ell x^\ell & c_{\ell+1}x^{\ell+1} & \cdots & c_{\ell+k-1}x^{\ell+k-1} \\ c_{\ell+1}x^{\ell+1} & \cdots & & \\ \vdots & & & \\ c_{\ell+k-1}x^{\ell+k-1} & \cdots & & c_{\ell+2k-2}x^{\ell+2k-2} \end{vmatrix} \neq 0$$

for $\ell = o,1,2,\ldots$ and $k=1,2,\ldots$
This determinant is a monomial of degree $k(\ell+k-1)$ in the variable x. Demanding that this monomial is nontrivial is equivalent with demanding that this determinant evaluated at $x = 1$ is nonzero. For a normal series we can construct a table with double entry of numbers $q_k^{(\ell)}$ and $e_k^{(\ell)}$ defined as follows:

$$e_o^{(\ell)} = 0 \qquad \ell = o,1,\ldots$$

$$q_1^{(\ell)} = \frac{c_{\ell+1}x^{\ell+1}}{c_\ell x^\ell} \qquad \ell = o,1,\ldots$$

$$e_k^{(\ell)} = q_k^{(\ell+1)} + e_{k-1}^{(\ell+1)} - q_k^{(\ell)} \qquad \ell = o,1,2,\ldots \qquad k = 1,2,\ldots$$

$$q_{k+1}^{(\ell)} = q_k^{(\ell+1)} e_k^{(\ell+1)} / e_k^{(\ell)} \qquad \ell = o,1,2,\ldots \qquad k = 1,2,\ldots$$

From this qd-algorithm we can obtain Padé approximants to the function F in the following way. The (n,m) Padé approximant for $n \geq m$ is equal to the $(2m)^{th}$ convergent K_{2m} of the continued fraction

$$c_o + c_1x + \ldots + c_{n-m}x^{n-m} + \frac{c_{n-m+1}\, x^{n-m+1}|}{|1} - \frac{q_1^{(n-m+1)}|}{|1} - \frac{e_1^{(n-m+1)}|}{|1} - \frac{q_2^{(n-m+1)}|}{|1} - \frac{e_2^{(n-m+1)}|}{|1} - \ldots ,$$

if $K_0 = \sum_{k=0}^{n-m} c_k x^k$, and it is also equal to the $(2m+1)^{th}$ convergent K_{2m+1} of the continued fraction

$$c_0+c_1x+\ldots+c_{n-m-1}x^{n-m-1}+\frac{c_{n-m}\,x^{n-m}|}{|1}-\frac{q_1^{(n-m)}|}{|1}-\frac{e_1^{(n-m)}|}{|1}-\frac{q_2^{(n-m)}|}{|1}-\frac{e_2^{(n-m)}|}{|1}-\ldots,$$

if $K_0 = \sum_{k=0}^{n-m-1} c_k x^k$ [8,9].

Both the terms $q_k^{(\ell)}$ and $e_k^{(\ell)}$ contain a factor x now because of the definition of $q_1^{(\ell)}$.

Let us now turn to the operator

$$F(x) = \sum_{k=0}^{\infty} C_k x^k$$

We call the series F normal if there exists x in X such that $H_k(C_{\ell+k-1}x^{\ell+k-1})$ is regular in Y for $\ell = 0,1,2,\ldots$ and $k=1,2,\ldots$

When the series

$$C_0 + \sum_{k=1}^{\infty} (C_k x^k - C_{k-1}x^{k-1})$$

is normal, then a representation of $(P(x),Q(x))$ satisfying (I.3.1) is given by (I.4.3) and (I.4.4). Normality of the series $C_0 + \sum_{k=1}^{\infty} (C_k x^k - C_{k-1}x^{k-1})$ is equivalent with $H_k(\Delta C_{\ell+k-1}x^{\ell+k-1})$ being regular in Y for some x in X. So normality of the series $C_0 + \sum_{k=0}^{\infty} \Delta C_k x^k$ implies regularity of $Q_{[n,m]}(x) = H_m(\Delta C_n x^n)$ and thus existence of $\frac{1}{Q_{[n,m]}} \cdot P_{[n,m]}$.

For a normal series F we can define the abstract qd-scheme as follows:

$$E_o^{(\ell)} = 0 \qquad \ell = o,1,\ldots$$

$$Q_1^{(\ell)} = (C_{\ell+1}x^{\ell+1}).(C_\ell\, x^\ell)^{-1} \qquad \ell = 0,1,\ldots$$

$$E_k^{(\ell)} = Q_k^{(\ell+1)} + E_{k-1}^{(\ell+1)} - Q_k^{(\ell)} \qquad \ell = 0,1,\ldots \quad k = 1,2,\ldots$$

$$Q_{k+1}^{(\ell)} = Q_k^{(\ell+1)} . E_k^{(\ell+1)} . (E_k^{(\ell)})^{-1} \qquad \ell = 0,1,\ldots \quad k = 1,2,\ldots$$

Let us construct the following continued fractions in the Banach algebra Y:

$$\sum_{k=0}^{n-m} C_k x^k + \cfrac{C_{n-m+1}\, x^{n-m+1}}{I - \cfrac{Q_1^{(n-m+1)}}{I - \cfrac{E_1^{(n-m+1)}}{I - \cfrac{Q_2^{(n-m+1)}}{I - \cfrac{E_2^{(n-m+1)}}{I - \dots}}}}} \tag{I.7.5}$$

and

$$\sum_{k=0}^{n-m-1} C_k x^k + \cfrac{C_{n-m}\, x^{n-m}}{I - \cfrac{Q_1^{(n-m)}}{I - \cfrac{E_1^{(n-m)}}{I - \cfrac{Q_2^{(n-m)}}{I - \cfrac{E_2^{(n-m)}}{I - \dots}}}}} \tag{I.7.6}$$

where dividing means multiplying by the inverse element for the multiplication in Y. We shall now prove that these continued fractions are of the same form as in the univariate case where only a factor x remains in $q_k^{(\ell)}$ and $e_k^{(\ell)}$ after division of numerator and denominator and we shall also prove that the convergents of these continued fractions yield our abstract Padé approximants.

Theorem I.7.3.:

If we write $Q_k^{(\ell)} = \frac{N_{q,k,\ell}}{D_{q,k,\ell}}$ and $E_k^{(\ell)} = \frac{N_{e,k,\ell}}{D_{e,k,\ell}}$ then $\partial N_{q,k,\ell} = \partial D_{q,k,\ell} + 1$

$\partial N_{e,k,\ell} = \partial D_{e,k,\ell} + 1$

Proof:

The proof is by induction.

For k = 1 we have

$$N_{q,k,\ell} = C_{\ell+1}\, x^{\ell+1} \text{ and } D_{q,k,\ell} = C_\ell x^\ell$$

$$N_{e,k,\ell} = (C_{\ell+1} x^{\ell+1})^2 - C_\ell x^\ell . C_{\ell+2} x^{\ell+2}$$

$$D_{e,k,\ell} = C_\ell x^\ell . C_{\ell+1} x^{\ell+1}$$

so that

$$\partial N_{q,k,\ell} = \ell+1 = \partial D_{q,k,\ell} + 1$$

$$\partial N_{e,k,\ell} = 2\ell+2 = \partial D_{e,k,\ell} + 1$$

Suppose the theorem holds for $Q_1^{(\ell)},\ldots,Q_k^{(\ell)},E_1^{(\ell)},\ldots,E_k^{(\ell)}$;
we shall prove it then for $Q_{k+1}^{(\ell)}$ and $E_{k+1}^{(\ell)}$.

Since $Q_{k+1}^{(\ell)} = Q_k^{(\ell+1)}.E_k^{(\ell+1)}.(E_k^{(\ell)})^{-1}$, we have

$$Q_{k+1}^{(\ell)} = \frac{N_{q,k,\ell+1}.N_{e,k,\ell+1}.D_{e,k,\ell}}{N_{e,k,\ell}.D_{q,k,\ell+1}.D_{e,k,\ell+1}} = \frac{N_{q,k+1,\ell}}{D_{q,k+1,\ell}}$$

Thus $\partial N_{q,k+1,\ell} = \partial N_{q,k,\ell+1} + \partial N_{e,k,\ell+1} + \partial D_{e,k,\ell} = \partial D_{q,k+1,\ell} + 1$

For $E_{k+1}^{(\ell)}$ the proof is analogous.

■

Consider now the following descending staircase:

$$P_{[n-m,0]}(x)\ \frac{1}{Q_{[n-m,0]}(x)}$$

$$P_{[n-m+1,0]}(x)\frac{1}{Q_{[n-m+1,0]}(x)} \qquad P_{[n-m+1,1]}(x)\ \frac{1}{Q_{[n-m+1,1]}(x)}$$

$$P_{[n-m+2,1]}(x)\ \frac{1}{Q_{[n-m+2,1]}(x)} \qquad \cdots$$

$$\vdots$$

Theorem I.7.4.:

$$P_{[n,m]}(x)\cdot\frac{1}{Q_{[n,m]}(x)} \text{ is the } (2m)^{th} \text{ convergent of the continued fraction (I.7.5)}$$

Proof:

Let $K_{i+j} = P_{[n-m+i,j]}(x)\ \frac{1}{Q_{[n-m+i,j]}(x)} \qquad i+j = 0,1,\ldots$

Regularity of the $H_k(C_\ell)$ and the $H_k(\Delta C_\ell)$ and the use of the formulas (I.7.1-4) imply that $K_{2i+1}-K_{2i}$, $K_{2i}-K_{2i-1}$, $K_{i+j}-K_{i+j-2}$ are regular.
So it is possible to construct the continued fraction

$$\sum_{k=o}^{n-m} C_k x^k + \cfrac{C_{n-m+1}\, x^{n-m+1}}{I + \cfrac{\frac{K_1-K_2}{K_2-K_o}}{I + \sum_{k=3}^{\infty} \left. \frac{(K_{k-1} - K_k)(K_{k-2}-K_{k-3})}{(K_k - K_{k-2})(K_{k-1}-K_{k-3})} \right| \overline{\left| \; I \right.}}} \tag{I.7.7}$$

with convergents $K_o, K_1, K_2, \ldots$ where dividing again means multiplying by the inverse element for the multiplication defined in Y.

It is easy to verify that

$$\frac{K_1-K_2}{K_2-K_o} = Q_1^{(n-m+1)} \text{ and } \frac{(K_2 - K_3)(K_1 - K_o)}{(K_3 - K_1)(K_2 - K_o)} = E_1^{(n-m+1)}$$

using the representation of $P_{[n-m,o]}(x)$, $Q_{[n-m,o]}(x)$, $P_{[n-m+1,o]}(x)$, $Q_{[n-m+1,o]}(x), \ldots$ given in § 4.

Let us denote

$$\frac{(K_{k-1} - K_k)(K_{k-2} - K_{k-3})}{(K_k - K_{k-2})(K_{k-1} - K_{k-3})}$$

by $A_{k/2}^{(n-m+1)}$ if k is even and by $B_{(k-1)/2}^{(n-m+1)}$ if k is odd. We write also $A_1^{(n-m+1)} = Q_1^{(n-m+1)}$.

If we write down the continued fraction that is the even contraction of (I.7.7) (i.e. a continued fraction having as convergents the K_{2k} for $k = o,1,2,\ldots$) we get

$$\sum_{k=o}^{n-m} C_k x^k + \cfrac{C_{n-m+1}\, x^{n-m+1}}{I - A_1^{(n-m+1)} - \cfrac{A_1^{(n-m+1)}\, B_1^{(n-m+1)}}{I - B_1^{(n-m+1)} - A_2^{(n-m+1)} - \ldots}} \tag{I.7.8}$$

If we write down the continued fraction that is the odd contraction of (I.7.7) with n-m replaced by n-m-1 (i.e. a continued fraction having as convergents the $P_{[n-m,o]}(x) \cdot \frac{1}{Q_{[n-m,o]}(x)}$, $P_{[n-m+1,1]}(x) \cdot \frac{1}{Q_{[n-m+1,1]}(x)}, \ldots$

on the descending staircase (I.7.10), we get

$$\cfrac{\sum_{k=o}^{n-m-1} C_k x^k + C_{n-m} x^{n-m} A_1^{(n-m)}}{I - A_1^{(n-m)} - B_1^{(n-m)} - \cfrac{B_1^{(n-m)} A_2^{(n-m)}}{I - A_2^{(n-m)} - B_2^{(n-m)} - \ldots}} \qquad (I.7.9)$$

Because (I.7.8) and (I.7.9) have the same convergents, we have

$$A_k^{(n-m+1)} B_k^{(n-m+1)} = B_k^{(n-m)} A_{k+1}^{(n-m)} \qquad k=1,2,\ldots$$

$$B_{k-1}^{(n-m+1)} + A_k^{(n-m+1)} = B_k^{(n-m)} + A_k^{(n-m)} \qquad k=1,2,\ldots$$

if we put $E_o^{(n-m+1)} = 0$.

So

$$A_k^{(n-m+1)} = Q_k^{(n-m+1)} \qquad k=1,2,\ldots$$

$$B_k^{(n-m+1)} = E_k^{(n-m+1)}$$

This completes the proof.

∎

Analogously we can formulate and prove the next theorem.

Theorem I.7.5.:

$P_{[n,m]}(x) \cdot \frac{1}{Q_{[n,m]}(x)}$ is the $(2m+1)^{th}$ convergent of the continued fraction (I.7.6).

This can easily be seen by writing down the continued fraction (I.7.7) with n-m replaced by n-m-1; the convergents of this continued fraction are the abstract rational operators on the following descending staircase:

$$P_{[n-m-1,o]}(x) \frac{1}{Q_{[n-m-1,o]}(x)}$$

$$P_{[n-m,o]}(x) \cdot \frac{1}{Q_{[n-m,o]}(x)} \qquad P_{[n-m,1]}(x) \cdot \frac{1}{Q_{[n-m,1]}(x)} \qquad (I.7.10)$$

$$P_{[n-m+1,1]}(x) \cdot \frac{1}{Q_{[n-m+1,1]}(x)} \qquad \ldots$$

$$\vdots$$

We illustrate the two preceding theorems by means of a simple example.
Again consider

$$F: C'([1,T]) \to C([1,T]): x(t) \to e^{x(t)} \frac{dx}{dt} - (1+c)$$

The unit in the Banach algebra $C([1,T])$ is the constant function $x(t)=1$; so we shall write $I = 1$. A representant of the (1,1) abstract Padé approximant is the second. convergent of the continued fraction (I.7.5):

$$\cfrac{-(1+c) + \frac{dx}{dt}}{1 - \cfrac{Q_1^{(1)}}{1}}$$

where $Q_1^{(1)} = x(t)$; it is also the third convergent of the continued fraction (I.7.6):

$$\cfrac{-(1+c)}{1 - \cfrac{Q_1^{(o)}}{1 - \cfrac{E_1^{(o)}}{1}}}$$

where $Q_1^{(o)} = -\frac{dx}{dt} / (1+c)$ and $E_1^{(o)} = Q_1^{(1)} - Q_1^{(o)}$. Indeed these convergents equal $\varepsilon_2^{(o)}$.

§ 8. EXISTENCE OF AN IRREDUCIBLE FORM

Let the couple of abstract polynomials $(P(x),Q(x))$ satisfy definition I.3.3. For some spaces X and Y a unique irreducible form $\frac{1}{Q_\star}.P_\star$ of the abstract rational operator $\frac{1}{Q}.P$ exists. We give some examples of such spaces X and Y.

a) For instance $X = \mathbb{R}^p$ or $\mathbb{C}^p$ and $Y = \mathbb{R}^q$ or $\mathbb{C}^q$ with a componentwise multiplication in Y, for every abstract polynomial $V : X \to Y$ with $D(V) \neq \emptyset$ has a unique prime factorization in the ring of abstract polynomials and thus an irreducible form $\frac{1}{Q_\star}.P_\star$ of $\frac{1}{Q}.P$ can be found by cancelling as many terms as possible in the unique prime factorization of P or Q. What's more, this irreducible form is unique now and all equivalent solutions (R, S) have the same irreducible form $\frac{1}{Q_\star}.P_\star$.

b) Consider a Banach algebra Z with unit I, not necessarily commutative. Take a, a regular element in Z.

Now $X = \{\lambda a \mid \lambda \in \Lambda\}$ is a Banach space and

$Y = \{\sum_{i=0}^{\infty} \lambda_i a^i \mid \lambda_i \in \Lambda\}$ is a commutative Banach algebra with unit I.

Every nonzero element of Y is regular, for

$y = \sum_{i=0}^{\infty} \lambda_i a^i$ can be written as $a^h \sum_{i=0}^{\infty} \lambda_{i+h} a^i$ with $\lambda_h \neq 0$ and

for every element $\sum_{i=0}^{\infty} \lambda_i a^i$ with $\lambda_0 \neq 0$, the inverse element for

the multiplication is $\sum_{j=0}^{\infty} \mu_j a^j$ with

$$\mu_j = (-1)^j \left[\frac{\lambda_1^j}{\lambda_0^{j+1}} + \sum_{k=2}^{j-1} (-1)^{k-1} \sum_{(i_1,\dots,i_k)\in I_k} \left[\frac{(j-k+1)!}{i_1!\dots i_k!} \frac{\lambda_1^{i_1} \dots \lambda_k^{i_k}}{\lambda_0^{j-k+2}} \right] + (-1)^{j-1} \frac{\lambda_j}{\lambda_0^2} \right]$$

where $I_k = \{(i_1,\dots,i_k) \in \mathbb{N}^k \mid i_1 + \dots + i_k = j-k+1 \text{ and } i_1.1 + \dots + i_k.k = j\}$
The ring of abstract polynomials is $Y[\lambda]$, the set of all polynomials in λ with coefficients in Y. Since Y is a field, $Y[\lambda]$ is a principal ideal domain [5 pp. 152] and thus every element in $Y[\lambda]$ has a unique prime factorization [5 pp. 155]; also every abstract polynomial V with $D(V) = \emptyset$ is identically 0.

Because of the unique prime factorization, an irreducible form $\frac{1}{Q_\star}.P_\star$ of $\frac{1}{Q}.P$ can be found and because there are no nontrivial abstract polynomials V with $D(V) = \emptyset$, this irreducible form $\frac{1}{Q_\star}.P_\star$ is unique and is also the irreducible form of $\frac{1}{S}.R$ where (R, S) is an equivalent solution of (I.3.1).

In the case of these two examples it is even true that the Banach algebra Y does not contain nilpotent elements. Then the only units V in the ring of abstract polynomials are the o-linear operators y with y a regular element in Y, because for a unit V we know that $\frac{1}{V}$ is also an abstract polynomial and $\partial V \leq \partial(V.\frac{1}{V}) = 0$ because of lemma I.2.3. For the sequel of this chapter we will restrict ourselves to Banach spaces X and Banach algebras Y without nilpotent elements such that a unique irreducible form of all solutions of (I.3.1) exists. Then $\partial_1 P_\star$, $\partial_1 Q_\star$ and $\partial_0 Q_\star$ do not depend anymore on the reduced rational form $\frac{1}{Q_\star}.P_\star$ we consider. We can now redefine the (n,m) abstract Padé approximant for an operator F.

Definition I.8.1.:

Let $(P(x), Q(x))$ satisfy definition I.3.3 and let

$D(P) \cup D(Q) \neq \emptyset$.

Let $\frac{1}{Q_\star}.P_\star$ be the irreducible form of $\frac{1}{Q}.P$.

a) If $Q_\star(0) = I$, we call $\frac{1}{Q_\star}.P_\star$ the <u>normalized (n,m) abstract Padé-approximant for F</u> (normalized (n,m) APA).

b) If $0 \notin D(Q_\star)$, we call $\frac{1}{Q_\star}.P_\star$ the <u>(n,m) abstract Padé-approximant for F</u> ((n,m) APA).

In definition I.8.1 a) the units are fixed by the normalization $Q_\star(0) = I$ and in definition I.8.1 b) the (n,m) APA is only unique up to units.
If for all the solutions (P,Q) of (I.3.1), $D(P) \cup D(Q) = \emptyset$, then we shall call the (n,m) APA undefined.
When a unique irreducible form of the solutions of (I.3.1) exists, more detailed information about the covariance of abstract Padé approximants can be given.
Let us take a look at the reciprocal covariance. If $0 \in D(F)$ and if in addition $D(T_{t_o}) \neq \emptyset$, then

$$0 \in D(Q_\star) \Rightarrow 0 \in D(P_\star)$$

because $\partial_o Q_\star = 0$ and $(C_o.B_{\star o}).T_{t_o} x^{t_o} = A_{\star o}.T_{t_o} x^{t_o}$, and also

$$0 \notin D(Q_\star) \Rightarrow 0 \notin D(P_\star)$$

because if $\partial_o Q_\star > 0$ also $\partial_o P_\star > 0$ (theorem I.5.1 b)) and if $B_{\star o}$ is not regular in Y also $A_{\star o}$ is not regular in Y $(C_o.B_{\star o} = A_{\star o})$. So the normalized (n,m) APA for F is transformed into the normalized (n,m) APA for $\frac{1}{F}$ and the (n,m) APA for F is transformed into the (n,m) APA for $\frac{1}{F}$, if $D(T_{t_o}) \neq \emptyset$.

Let us take a look at the covariance property I.6.2. If $0 \in D(c.F+d)$ and if in addition $(a.d-b.c)$ is regular in Y, then $\frac{1}{(c.P_\star+d.Q_\star)}.(a.P_\star+b.Q_\star)$ is the irreducible form of $\frac{1}{(c.P+d.Q)}.(a.P+b.Q)$. If $D(T_{t_o}) \neq \emptyset$, also

$$0 \in D(Q_\star) \Rightarrow 0 \in D(c.P_\star+d.Q_\star)$$

because $(c.P_\star+d.Q_\star)(0) = (c.C_o+d).B_{\star o}$, and

$$0 \notin D(Q_\star) \Rightarrow 0 \notin D(c.P_\star+d.Q_\star)$$

because for $\partial_o Q_\star > 0$ also $\partial_o P_\star > 0$ (theorem I.5.1 b)) and if $B_{\star o}$ is not regular in Y also $(c.P_\star+d.Q_\star)(0) = c.A_{\star o}+d.B_{\star o} = (c.C_o+d).B_{\star o}$ is not regular in Y. So the normalized (n,m) APA for F is transformed into the normalized (n,m) APA for $\frac{1}{c.F+d}.(a.F+b)$ and the (n,m) APA for F is transformed into the (n,m) APA for $\frac{1}{c.F+d}.(a.F+b)$, if $D(T_{t_o}) \neq \emptyset$ and $(a.d-b.c)$ is regular in Y.

Let us take a look at the covariance property I.6.3. Because here $S_\star(0) = Q_\star(0)$, automatically the normalized (n,m) APA is transformed into the normalized (n,m) APA and the (n,m) APA is transformed into the (n,m) APA.
From now on we shall often consider the normalized (n,m) APA to be a special case of the (n,m) APA and not mention the specification normalized.

§ 9. FINITE DIMENSIONAL SPACES

a) When $X = \mathbb{R} = Y$ ($\Lambda = \mathbb{R}$), then the definition of abstract Padé approximant is precisely the classical definition of univariate Padé approximant. F is now a real-valued function of one real variable, with a Taylor series development $\sum_{k=o}^{\infty} c_k x^k$ where $c_k = \frac{1}{k!} F^{(k)}(0)$ is a real number.
The k-linear and bounded operators C_k are $C_k x^k = c_k . \underbrace{x \ldots x}_{k}$

The j-linear and bounded operators $B_j x^j = b_j . \underbrace{x \ldots x}_{j}$ for $j=nm,\ldots,nm+m$ such that

$$\begin{cases} c_{n+1} . b_{nm} + \ldots + c_{n+1-m} . b_{nm+m} = 0 \\ \vdots \\ c_{n+m} . b_{nm} + \ldots + c_n . b_{nm+m} = 0 \end{cases}$$

are a solution of (I.4.2).
The i-linear and bounded operators $A_i x^i = a_i . \underbrace{x \ldots x}_{i}$ for $i=nm,\ldots nm+n$ such that

$$\begin{cases} c_o . b_{nm} = a_{nm} \\ c_1 . b_{nm} + c_o . b_{nm+1} = a_{nm+1} \\ \vdots \\ c_n . b_{nm} + \ldots + c_o . b_{nm+n} = a_{nm+n} \end{cases}$$

are a solution of (I.4.1).
The irreducible form

$$\frac{1}{Q_\star(x)} . P_\star(x) \text{ of } \frac{1}{Q(x)} . P(x) = \frac{1}{\sum_{j=nm}^{nm+m} b_j x^j} . \sum_{i=nm}^{nm+n} a_i x^i \text{ such that } Q_\star(0) = 1,$$

is also the irreducible form $\frac{1}{Q_\star(x)} . P_\star(x)$ of $\dfrac{\sum_{i=nm}^{nm+n} a_i x^{i-nm}}{\sum_{j=nm}^{nm+m} b_j x^{j-nm}}$ with $Q_\star(0) = 1$.

b) If $X = \mathbb{R}^p$ and $Y = \mathbb{R}$ $(\Lambda = \mathbb{R})$, then F is a real-valued function of p real variables. Now $L(X^k,Y)$ is isomorphic with $\mathbb{R}^{p^k}$. More about multivariate Padé approximants can be found in chapter II.

c) If $X = \mathbb{R}^p$ and $Y = \mathbb{R}^q$ $(\Lambda = \mathbb{R})$, then F is a system of q real-valued functions in p real variables. Now $L(X,Y)$ is isomorphic with $\mathbb{R}^{q\times p}$ and $L(X^k,Y)$ isomorphic with $\mathbb{R}^{q\times p^k}$. An element of $\mathbb{R}^{q\times p^k}$ is represented by a row of p^{k-1} matrices (blocks), each containing q rows and p columns [41]. For a k-linear and bounded operator $C_k = (c_{i_1 \ldots i_{k+1}})$ we have

i_1 = row-index in the block

$i_2 \ldots i_k$ = number of the block (the most right index grows the fastest)

i_{k+1} = column-index in the block.

So $C_k = (c_{i_1 \ldots i_{k+1}})$ looks like

$$\left(\cdots \left| \begin{matrix} c_{1i_2\ldots i_k1} & \cdots & c_{1i_2\ldots i_kp} \\ c_{qi_2\ldots i_k1} & \cdots & c_{qi_2\ldots i_kp} \end{matrix} \right| \cdots \right)$$

and an evaluation $(c_{i_1 \ldots i_{k+1}}) \begin{pmatrix} x_1 \\ \vdots \\ x_p \end{pmatrix}$ is performed like

$$\left(\cdots \left| \begin{matrix} \sum_{j=1}^{p} c_{1i_2\ldots i_kj}\, x_j \\ \vdots \\ \sum_{j=1}^{p} c_{qi_2\ldots i_kj}\, x_j \end{matrix} \right| \cdots \right)$$

Thus the result of one evaluation is a hypermatrix containing q rows of p^{k-1} numbers, i.e. a row of p^{k-2} matrices (blocks) each contaning q rows and p columns; in other words the result of one evaluation is a (k-1)-linear and bounded operator.
The abstract polynomials (P,Q) satisfying definition I.3.3 now have for each of the q components the form of the multivariate polynomials in b). More about the solution of a system of nonlinear equations in p real variables by means of abstract Padé approximants, can be found in chapter III.

§ 10. THE ABSTRACT PADE TABLE

Let $R_{n,m}$ denote the (normalized) (n,m) APA for F if it is not undefined. The $R_{n,m}$ can be ordered for different values of n and m in a table:

$$\begin{array}{cccc} R_{o,o} & R_{o,1} & R_{o,2} & \cdots \\ R_{1,o} & R_{1,1} & R_{1,2} & \cdots \\ R_{2,o} & R_{2,1} & \cdots & \\ R_{3,o} & \vdots & & \\ \vdots & & & \end{array}$$

Gaps can occur in this table because of undefined elements.

We will now prove that the abstract Padé table consists of squares of equal elements under the following condition, numbered (I.10.1).

Let (P,Q) be a solution of (I.3.1). Let $R_{n,m} = \frac{1}{Q_\star}.P_\star$ and $T(x) = \sum_{k=t_o}^{\partial T} T_k x^k$.

We need a solution (P,Q) where

$$D(T_{t_o}) \neq \emptyset \qquad (I.10.1)$$

to be able to prove the block-structure of the Padé table. For every (n,m) where this condition is not satisfied the block-structure may be disturbed. An example of this phenomenon will be given after theorem I.10.1 . First of all we shall prove the following lemma which we shall frequently use in the next proofs.

Lemma I.10.1.:

Take x_o in X, $x_o \neq 0$. For every n in $\mathbb{N}$, there exists D_n in $L(X^n,Y)$ such that $D_n x_o^n = I$.

Proof:

Let n = 1.

Take x_o in X, $x_o \neq 0$ and define the linear functional [41 pp. 34]

$f : M = \{\lambda x_o \mid \lambda \in \Lambda\} \to \Lambda : \lambda x_o \to \lambda.$

Now $|f(\lambda x_o)| = |\lambda| = \frac{\|\lambda x_o\|}{\|x_o\|}$.

Define the norm $p(x) = \frac{\|x\|}{\|x_o\|}$ on X. So $|f(x)| \leq p(x)$ for every x in M.

By the functional analysis theorem of Hahn-Banach [43 pp. 57]

this linear functional f can be extended to a linear functional

$\tilde{f} : X \to \Lambda$ such that $\tilde{f}(x) = f(x)$ for every x in M and such that $|\tilde{f}(x)| \le p(x)$ for every x in X.

We now define $D_1 : X \to Y : x \to \tilde{f}(x).I$.

Clearly $D_1 \in L(X,Y)$ and $D_1\, x_o = I$ since $\tilde{f}(x_o) = f(x_o) = 1$.

If $D_{n-1} \in L(X^{n-1},Y)$ with $D_{n-1}\, x_o^{n-1} = I$ then we can define for x in X : $D_n\, x = \tilde{f}(x).D_{n-1}$ and so $D_n \in L(X^n,Y)$.

Now $D_n x_o^n = \tilde{f}(x_o).D_{n-1}\, x_o^{n-1} = I$.

■

This lemma implies that for $x_o \neq 0$ and n given, we can always find D_n in $L(X^n,Y)$ with $x_o \in D(D_n)$.

Theorem I.10.1.:

Let $\frac{1}{Q_\star}.P_\star = R_{n,m}$ for F. Let (I.10.1) be satisfied.

Then a) $\partial_o(F.Q_\star - P_\star) = \partial_o Q_\star + \partial_1 P_\star + \partial_1 Q_\star + t + 1$ with $t \ge o$

b) $n \le \partial_1 P_\star + t$ and $m \le \partial_1 Q_\star + t$

c) if $\partial_o Q_\star \le \partial_1 P_\star . \partial_1 Q_\star$ then for all integers i,j satisfying $\partial_1 P_\star \le i \le \partial_1 P_\star + t$ and $\partial_1 Q_\star \le j \le \partial_1 Q_\star + t$: $R_{i,j} = R_{n,m}$.

Proof:

a) In theorem I.5.4 b) we proved that

$$(F.Q_\star - P_\star)(x) = O(x^{\partial_o Q_\star + \partial_1 P_\star + \partial_1 Q_\star + 1}),$$

in other words that $\partial_o(F.Q_\star - P_\star) \ge \partial_o Q_\star + \partial_1 P_\star + \partial_1 Q_\star + 1$.

Write $\partial_o(F.Q_\star - P_\star) = \partial_o Q_\star + \partial_1 P_\star + \partial_1 Q_\star + t + 1$ with $t \ge o$.

b) Suppose $n > \partial_1 P_\star + t$ or $m > \partial_1 Q_\star + t$.

Then for every r, $o \le r \le \min(n - \partial_1 P_\star,\ m - \partial_1 Q_\star)$ and for every $T_{nm-\partial_o Q_\star + r}$ in $L(X^{nm-\partial_o Q_\star + r},Y)$ with

$D(T_{nm-\partial_o Q_\star + r}) \neq \emptyset$ we have

$$\partial_o[T_{nm-\partial_o Q_\star + r}.(F.Q_\star - P_\star)] = (nm-\partial_o Q_\star + r) + (\partial_o Q_\star + \partial_1 P_\star + \partial_1 Q_\star + t + 1) < nm+n+m+$$

This is in contradiction with theorem I.5.3.

c) Let $s = \min(i - \partial_1 P_\star, j - \partial_1 Q_\star)$.

Since $\partial_0 Q_\star \le \partial_1 P_\star . \partial_1 Q_\star$, we know that $i.j - \partial_0 Q_\star + s \ge o$.

Take D_s in $L(X^{i.j-\partial_0 Q_\star + s}, Y)$ with

$D(D_s) \cap (D(P_\star) \cup D(Q_\star)) \ne \emptyset$

which is possible because of lemma I.10.1.

For $P_1 = P_\star . D_s$ and $Q_1 = Q_\star . D_s$ we have

$$\begin{cases} \partial_0 P_1 \ge \partial_0 P_\star + (i.j - \partial_0 Q_\star + s) \ge i.j \\ \partial_0 Q_1 \ge \partial_0 Q_\star + (i.j - \partial_0 Q_\star + s) \ge i.j \end{cases}$$

$$\begin{cases} \partial P_1 \le (\partial_1 P_\star + \partial_0 Q_\star) + (i.j - \partial_0 Q_\star + s) \le i.j + i \\ \partial Q_1 \le (\partial_1 Q_\star + \partial_0 Q_\star) + (i.j - \partial_0 Q_\star + s) \le i.j + j \end{cases}$$

and $(F.Q_1 - P_1)(x) = O(x^{i.j+\partial_1 P_\star + \partial_1 Q_\star + s+t+1})$

Since $i \le \partial_1 P_\star + t$ and $j \le \partial_1 Q_\star + t$ we know that

$i.j + i + j + 1 \le i.j + \partial_1 P_\star + \partial_1 Q_\star + s + t + 1$.

So $(F.Q_1 - P_1)(x) = O(x^{i.j+i+j+1})$.

∎

Remark the fact that if one element of a square in the abstract Padé table is defined, all the elements of the same square are because of the constructive proof of theorem I.10.1 c). Also if one element of a square is a normalized APA, then all the elements of the same square are.

We now give an example where the block-structure is disturbed because (I.10.1) is not satisfied.

$$\text{Take } F : \mathbb{R}^2 \to \mathbb{R}^2 : \binom{x}{y} \to \begin{pmatrix} 1 + \dfrac{y}{1+y^2} \\ 1 - \cos x \end{pmatrix} = \begin{pmatrix} 1 + \sum_{k=o}^{\infty} (-1)^k y^{2k+1} \\ \sum_{k=1}^{\infty} \dfrac{(-1)^{k+1}}{(2k)!} x^{2k} \end{pmatrix}$$

Now for all $h > o$: $R_{h,1} = R_{h,o}$. We shall explain this.

For h even:

$$B_h \binom{x}{y}^h = \begin{pmatrix} 0 \\ (-1)^{\frac{h}{2}} \dfrac{x^h}{h!} \end{pmatrix}, \qquad B_{h+1} \binom{x}{y}^{h+1} = \begin{pmatrix} (-1)^{\frac{h}{2}} y^{h+1} \\ 0 \end{pmatrix},$$

$$A_h \binom{x}{y}^h = 0 \qquad , \qquad A_{h+1} \binom{x}{y}^{h+1} = \begin{pmatrix} (-1)^{\frac{h}{2}} y^{h+1} \\ 0 \end{pmatrix}$$

and

$$A_{h+\ell} \binom{x}{y}^{h+\ell} = \begin{pmatrix} (-1)^{\frac{h+\ell+2}{2}} y^{h+\ell} \\ (-1)^{\frac{h+\ell+2}{2}} \frac{x^{h+\ell}}{h!\ell!} \end{pmatrix} \qquad \text{for } \ell \text{ even and } 2 \le \ell \le h$$

$$A_{h+\ell} \binom{x}{y}^{h+\ell} = 0 \quad \text{for } \ell \text{ odd and } 3 \le \ell < h$$

are a solution of (I.4.1) and (I.4.2).

For h odd:

$$B_h \binom{x}{y}^h = \begin{pmatrix} (-1)^{\frac{h-1}{2}} y^h \\ 0 \end{pmatrix} , \qquad B_{h+1} \binom{x}{y}^{h+1} = \begin{pmatrix} 0 \\ (-1)^{\frac{h+3}{2}} \frac{x^{h+1}}{(h+1)!} \end{pmatrix} ,$$

$$A_h \binom{x}{y}^h = \begin{pmatrix} (-1)^{\frac{h-1}{2}} y^h \\ 0 \end{pmatrix} , \qquad A_{h+1} \binom{x}{y}^{h+1} = \begin{pmatrix} (-1)^{\frac{h-1}{2}} y^{h+1} \\ 0 \end{pmatrix}$$

and

$$A_{h+\ell} \binom{x}{y}^{h+\ell} = 0 \quad \text{for } \ell \text{ even and } 2 \le \ell < h$$

$$A_{h+\ell} \binom{x}{y}^{h+\ell} = \begin{pmatrix} (-1)^{\frac{h+\ell-2}{2}} y^{h+\ell} \\ (-1)^{\frac{h+\ell}{2}} \frac{x^{h+\ell}}{(h+1)!(\ell-1)!} \end{pmatrix} \qquad \text{for } \ell \text{ odd and } 3 \le \ell \le h$$

are a solution of (I.4.1) and (I.4.2).

For all h numerator and denominator of the solution have been devided by $B_h\binom{x}{y}^h + B_{h+1}\binom{x}{y}^{h+1}$ to get the irreducible form $R_{h,1}$. Now $D(B_h) = \emptyset$ and $D(B_{h+1}) = \emptyset$ and we cannot find T_h in $L(X^h,Y)$ or T_{h+1} in $L(X^{h+1},Y)$ such that:

$(P_\star.T_h, Q_\star.T_h)$ satisfies (I.4.1) and (I.4.2) and $D(T_h) \neq \emptyset$

or

$(P_\star.T_{h+1}, Q_\star.T_{h+1})$ satisfies (I.4.1) and (I.4.2) and $D(T_{h+1}) \neq \emptyset$

So (I.10.1) is not satisfied for the normalized (h,1) APA. Other examples where (I.10.1) is not satisfied will be discussed in the next paragraph.

§ 11. REGULARITY AND NORMALITY

11.1. *Definitions*

Regularity and normality are also defined exactly as in the case of univariate Padé approximants.

Definition I.11.1.:

The (n,m) APA $\frac{1}{Q_\star}.P_\star$ for F is called <u>regular</u> if

$$(F.Q_\star - P_\star)(x) = O(x^{\partial_o Q_\star + n+m+t+1}) \text{ with } t \geq o.$$

Definition I.11.2.:

The (n,m) APA $\frac{1}{Q_\star}.P_\star$ for F is called <u>normal</u> if it occurs only once in the abstract Padé-table.

Clearly the elements in the first column of the Padé table are regular because the (n,o) APA is the nth partial sum of the Taylor series development of F and $(F.Q-P)(x) = O(x^{n+1})$. If $C_o = F(0)$ is regular in Y then also the first row of the Padé table consists of regular abstract Padé approximants.

11.2. *Normality*

The following theorem makes clear that, under the assumption of (I.10.1), normality is stronger than regularity.

Theorem I.11.1.:

Let (I.10.1) be satisfied and let $\partial_o Q_\star \le \partial_1 P_\star . \partial_1 Q_\star$. The (n,m) APA $\frac{1}{Q_\star}.P_\star$ for F is normal if and only if $\partial_1 P_\star = n$ and $\partial_1 Q_\star = m$ and $\partial_o(F.Q_\star - P_\star) = \partial_o Q_\star + n+m+1$.

Proof: $\Rightarrow$

Since $R_{n,m}$ is normal, t = o in theorem I.10.1 c).

According to theorem I.10.1 b) we have $n \le \partial_1 P_\star$ and $m \le \partial_1 Q_\star$.

Because of theorem I.5.2 b) we also have $\partial_1 P_\star \le n$ and $\partial_1 Q_\star \le m$.

So $n = \partial_1 P_\star$ and $m = \partial_1 Q_\star$.

According to theorem I.10.1 a) we then have

$\partial_o(F.Q_\star - P_\star) = \partial_o Q_\star + n+m+1$

$\Leftarrow$

The proof goes by contraposition.

Suppose we can find i,j with i > n or j > m and such that $R_{i,j} = R_{n,m}$ (because of theorem I.10.1 c) we have in any case that $n \le i$ and $m \le j$). For every integer s and for every D_s in $L(X^{i.j+s-\partial_o Q_\star}, Y)$ with $D(D_s) \cap (D(P_\star) \cup D(Q_\star)) \neq \emptyset$ and with

$[(F.Q_\star - P_\star).D_s](x) = O(x^{i.j+i+j+1})$, we have

$i.j+i+j+1 \le i.j+n+m+s+1$

because $\partial_o [(F.Q_\star - P_\star).D_s] = i.j+n+m+s+1$.

So s > i-n or s > j-m. This is in contradiction with theorem I.5.3.

∎

Because of the formulas (I.4.3) and (I.4.4) we have for $P_{[n,m]}$ and $Q_{[n,m]}$:

$$B_{n.m+m}\, x^{n.m+m} = (-1)^m H_m (C_{n-m+2}\, x^{n-m+2})$$

$$A_{n.m+n}\, x^{n.m+n} = H_{m+1} (C_{n-m}\, x^{n-m})$$

$$B_{n.m}\, x^{n.m} = H_m (C_{n-m+1}\, x^{n-m+1})$$

$$(F.Q_{[n,m]} - P_{[n,m]})(x) = (-1)^m H_{m+1} (C_{n-m+1}\, x^{n-m+1}) + \dots$$

The following theorem is also a generalization of a well-known classical result which states that normality is equivalent with the non-triviality of 4 determinants.

Theorem I.11.2.:

Let $\partial_0 Q_\star \le \partial_1 P_\star . \partial_1 Q_\star$. If $D(P_{[n,m]}) \cup D(Q_{[n,m]}) \ne \emptyset$ and if $T(x) = T_{nm-\partial_0 Q_\star} x^{nm-\partial_0 Q_\star}$ then the (n,m) APA $\frac{1}{Q_\star}.P_\star$ for F is normal if and only if

$H_m(C_{n+1-m}x^{n+1-m}) \not\equiv 0$

$H_m(C_{n+2-m}x^{n+2-m}) \not\equiv 0$

$H_{m+1}(C_{n-m}x^{n-m}) \not\equiv 0$

$H_{m+1}(C_{n+1-m}x^{n+1-m}) \not\equiv 0$

Proof: $\Rightarrow$

Since $Q_{[n,m]} = Q_\star.T$ we have

$\partial_0 Q = \partial_0 Q_\star + t_0 = \partial_0 Q_\star + (nm - \partial_0 Q_\star) = nm$ and so $H_m(C_{n+1-m}x^{n+1-m}) \not\equiv 0$.

Suppose $H_m(C_{n+2-m}x^{n+2-m}) \equiv 0$.

$$\begin{aligned} \text{Then } \partial_1 Q_\star &= \partial Q_\star - \partial_0 Q_\star \\ &\le \partial Q_{[n,m]} - t_0 - \partial_0 Q_\star \quad \text{because of lemma I.2.3} \\ &< nm+m- t_0 - \partial_0 Q_\star \\ &= m \text{ because } t_0 = nm - \partial_0 Q_\star . \end{aligned}$$

Suppose $H_{m+1}(C_{n-m}x^{n-m}) \equiv 0$

$$\begin{aligned} \text{Then } \partial_1 P_\star &= \partial P_\star - \partial_0 Q_\star \\ &\le \partial P_{[n,m]} - t_0 - \partial_0 Q_\star \quad \text{because of lemma I.2.3} \\ &< nm+n- t_0 - \partial_0 Q_\star \\ &= n \text{ because } t_0 = nm - \partial_0 Q_\star . \end{aligned}$$

These conclusions contradict the normality of $\frac{1}{Q_\star}.P_\star$

Because $D(T) \ne \emptyset$ we have $\partial_0(F.Q_{[n,m]} - P_{[n,m]}) = \partial_0(F.Q_\star - P_\star) + (nm - \partial_0 Q_\star)$

$= nm+n+m+1$

and so $H_{m+1}(C_{n+1-m}x^{n+1-m}) \not\equiv 0$.

Since $\partial P_{[n,m]} = nm+n$ and $P_{[n,m]} = P_\star . T$ we have $\partial P_\star = nm+n-(nm-\partial_0 Q_\star)$ and so $\partial_1 P_\star = n$.

Since $\partial Q_{[n,m]} = nm+m$ and $Q_{[n,m]} = Q_\star . T$ we have $\partial Q_\star = nm+m-(nm-\partial_0 Q_\star)$ and so $\partial_1 Q_\star = m$.

Because $D(T) \neq \emptyset$ we have
$$\partial_0(F.Q_\star - P_\star) = \partial_0(F.Q_{[n,m]} - P_{[n,m]}) - (nm-\partial_0 Q_\star)$$
$$= \partial_0 Q_\star + n+m+1.$$

So $\frac{1}{Q_\star} . P_\star$ is normal.

■

3. *Regularity*

The following theorem is a criterion for regularity.

Theorem I.11.3.:

The (n,m) APA $\frac{1}{Q_\star} . P_\star$ for F is regular if (I.10.1) is satisfied with $t_0 = nm-\partial_0 Q_\star$.

Proof:

Since $(P,Q) = (P_\star . T,\ Q_\star . T)$ satisfies (I.4.1) and (I.4.2), we have $\partial_0(F.Q-P) \geq n.m+n+m+1$.

Because $D(T_{nm-\partial_0 Q_\star}) \neq \emptyset$, we can conclude that
$$\partial_0(F.Q_\star - P_\star) = \partial_0(F.Q-P) - (nm-\partial_0 Q_\star) \geq \partial_0 Q_\star + n+m+1.$$

■

The following example illustrates that if the (n,m) APA is regular, we do not necessarily have that $D(T_{t_0}) \neq \emptyset$ or $t_0 = nm-\partial_0 Q_\star$.

Consider

$$F : \mathbb{R}^2 \to \mathbb{R}^2 : \binom{x}{y} \to \begin{pmatrix} \dfrac{xe^x - ye^y}{x-y} \\ 1 + \dfrac{x}{0.1-y} + \sin(xy) \end{pmatrix} = \begin{pmatrix} \sum_{i,j=0}^{\infty} \dfrac{1}{(i+j)!} x^i y^j \\ 1 + \sum_{k=0}^{\infty} \left(10^{k+1} x y^k + (-1)^k \dfrac{(xy)^{2k+1}}{(2k+1)!}\right) \end{pmatrix}$$

The (1,1) APA is
$$\begin{pmatrix} \dfrac{x+y+0.5(x^2+3xy+y^2)}{x+y-0.5(x^2+xy+y^2)} \\ \dfrac{1+10x-10.1y}{1-10.1y} \end{pmatrix}.$$

The (1,1) APA is regular since $\partial_o(F.Q_\star - P_\star) = 3 = \partial_o Q_\star + n+m+1$ with $\partial_o Q_\star = o$.
Now $T\binom{x}{y} = \binom{1}{10x}$. So $T_{t_o}\binom{x}{y}^{t_o} = \binom{1}{0}$ with $t_o = o$.

11.4. *Numerical examples*

Let us now illustrate these results by some numerical examples. Take

$$F: \mathbb{R}^2 \to \mathbb{R}: \binom{x}{y} \to 1 + \frac{x}{0.1-y} + \sin(xy)$$
$$= 1+10x+101xy + \sum_{k=3}^{\infty} 10^k xy^{k-1} + \sum_{k=1}^{\infty} (-1)^k \frac{(xy)^{2k+1}}{(2k+1)!}$$

Clearly $D(T_{t_o}) \neq \emptyset$ for all $R_{n,m}$.

The (1,1) APA is $\frac{1+10x-10.1y}{1-10.1y}$ with

$\partial_o Q_\star = o$, $\partial_1 P_\star = 1$, $\partial_1 Q_\star = 1$, $\partial_o(F.Q_\star - P_\star) = 3$.

So it is a normal element in the abstract Padé-table.

The (3,1) APA is $\frac{1+10x-10y+xy-10xy^2}{1-10y}$ with

$\partial_o Q_\star = o$, $\partial_1 P_\star = 3$, $\partial_1 Q_\star = 1$, $\partial_o(F.Q_\star - P_\star) = 6$ and $t_o = 3$.

So it is a regular element in the abstract Padé-table and we have the following square of equal elements: $R_{3,1} = R_{3,2} = R_{4,1} = R_{4,2}$.

The (1,2) APA is $\frac{x-1.01y+10y^2+10x^2-20.2xy}{x-1.01y+10y^2-10.1xy+2.01xy^2}$ with

$\partial_o Q_\star = 1$, $\partial_1 P_\star = 1$, $\partial_1 Q_\star = 2$, $\partial_o(F.Q_\star - P_\star) = 5$.

So it is also a normal element.

The (3,3) APA is

$$\frac{y + \frac{201}{6}.10^{-5}x^2 + 10y(x-y) + xy^2(1-10y) + \frac{1}{600}x^2(2.01x-y) + \frac{1}{60}x^2y(1.0301x-y)}{y + \frac{201}{6}.10^{-5}x^2 - 10y^2 - \frac{1}{600}x^2y - \frac{1}{60}x^2y^2}$$

with $\partial_o Q_\star = 1$, $\partial_1 P_\star = 3$, $\partial_1 Q_\star = 3$, $\partial_o(F.Q_\star - P_\star) = 8$ and thus it is also a normal element in the abstract Padé-table.

§ 12. PROJECTION PROPERTY AND PRODUCT PROPERTY

Consider Banach spaces X_i, $i=1,\dots,p < \infty$.

The space $X = \prod_{i=1}^{p} X_i$, normed by one of the following Minkowski-norms

$$\|x\|_q = \left(\sum_{i=1}^{p} \|x_i\|_{(i)}^q\right)^{1/q}$$

$$\|x\|_1 = \sum_{i=1}^{p} \|x_i\|_{(i)}$$

$$\|x\|_\infty = \max(\|x_1\|_{(1)}, \ldots, \|x_p\|_{(p)})$$

where $\|x_i\|_{(i)}$ is the norm of x_i in X_i and $x = (x_1,\ldots,x_p)$, is also a Banach space. We introduce the following notations

$${}^j\tilde{x} = (x_1, \ldots, x_{j-1}, 0, x_{j+1}, \ldots, x_p)$$

$$x_{'j'} = (x_1, \ldots, x_{j-1}, x_{j+1}, \ldots, x_p)$$

Theorem I.12.1.:

Let $X = \prod_{i=1}^{p} X_i$ and $\frac{1}{Q_\star}.P_\star$ be the (n,m) APA for $F: X \to Y$ and $j \in \{1,\ldots,p\}$. Let (I.10.1) be satisfied.

If $S(x_{'j'}) := Q_\star({}^j\tilde{x})$

$R(x_{'j'}) := P_\star({}^j\tilde{x})$

$G_j(x_{'j'}) := F({}^j\tilde{x})$

$D(S) \cup D(R) \neq \emptyset$

then the irreducible form $\frac{1}{S_\star}.R_\star$ of $\frac{1}{S}.R$ is the (n,m) APA for G_j.

Proof:

First we remark that for a bounded k-linear operator L of $L(X^k,Y)$, if the operator M is defined by

$Mx_{'j'}^{\ k} = M(x_1, \ldots, x_{j-1}, x_{j+1}, \ldots, x_p)^k := L\,{}^j\tilde{x}^k$, then

M is a bounded k-linear operator of $L((\prod_{\substack{i=1\\ i\neq j}}^{p} X_i)^k, Y)$.

Because (I.10.1) is satisfied, we have $t \geq o$ such that

$$\partial_o(F.Q_\star - P_\star) = \partial_o Q_\star + \partial_1 P_\star + \partial_1 Q_\star + t + 1$$

$$\partial_1 P_\star \leq n \leq \partial_1 P_\star + t$$

$$\partial_1 Q_\star \leq m \leq \partial_1 Q_\star + t$$

Using one of the Minkowski-norms $\|\ \|_q$ $(1 \le q \le \infty)$,

$\|{}^j\tilde{x}\|_q = \|(x_1, \ldots, x_{j-1}, 0, x_{j+1}, \ldots, x_p)\|_q$ in $\prod_{i=1}^{p} X_i$ equals

$\|x_{,j,}\|_q = \|(x_1, \ldots, x_{j-1}, x_{j+1}, \ldots, x_p)\|_q$ in $\prod_{\substack{i=1\\i\neq j}}^{p} X_i$.

Thus $(F.Q_\star - P_\star)({}^j\tilde{x}) = (G_j.S-R)(x_{,j,}) = O(x_{,j,}^{\partial_0 Q_\star + \partial_1 P_\star + \partial_1 Q_\star + t + 1})$

Now $\partial P_\star = \partial_1 P_\star + \partial_0 Q_\star \le \partial(P_\star.T) - t_0 \le nm+n$

and $\partial Q_\star = \partial_1 Q_\star + \partial_0 Q_\star \le \partial(Q_\star.T) - t_0 \le nm+m$

and so $s = nm - \partial_0 Q_\star + \min(n - \partial_1 P_\star, m - \partial_1 Q_\star) \ge o$.

Take D_s in $L((\prod_{\substack{i=1\\i\neq j}}^{p} X_i)^s, Y)$ with $D(D_s) \cap (D(S) \cup D(R)) \neq \emptyset$.

Then $\partial_0(R.D_s) \ge nm$, $\partial_0(S.D_s) \ge nm$

$$\partial(R.D_s) \le \partial_0 Q_\star + \partial_1 P_\star + nm - \partial_0 Q_\star + \min(n - \partial_1 P_\star, m - \partial_1 Q_\star) \le nm + n$$

$$\partial(S.D_s) \le \partial_0 Q_\star + \partial_1 Q_\star + nm - \partial_0 Q_\star + \min(n - \partial_1 P_\star, m - \partial_1 Q_\star) \le nm + m$$

$$[(G_j.S-R).D_s](x_{,j,})$$

$$= O\,(x_{,j,}^{\partial_0 Q_\star + \partial_1 P_\star + \partial_1 Q_\star + t + \min(n-\partial_1 P_\star, m-\partial_1 Q_\star) + nm - \partial_0 Q_\star + 1})$$

$= O\,(x_{,j,}^{nm+n+m+1})$ because $m \le \partial_1 Q_\star + t$ and

$n \le \partial_1 P_\star + t$

The irreducible form of $\frac{1}{S.D_s}.(R.D_s)$ is also the irreducible form of $\frac{1}{S}.R$ and this terminates the proof.

∎

We now return to the situation where the (n,m) abstract Padé approximant is an equivalence class.

First we searched for a product property of the following kind. Let X_1, X_2 be Banach spaces and Y a commutative Banach algebra. If $(P_{\star 1}(x_1), Q_{\star 1}(x_1))$ is the (n,m) abstract Padé approximant for the operator $F_1: X_1 \to Y$ and $(P_{\star 2}(x_2), Q_{\star 2}(x_2))$ is the (n,m) abstract Padé approximant for the operator $F_2: X_2 \to Y$, is then $(P_\star(x_1,x_2), Q_\star(x_1,x_2)) = (P_{\star 1}(x_1).P_{\star 2}(x_2), Q_{\star 1}(x_1).Q_{\star 2}(x_2))$ the (n,m) abstract Padé approximant for $F: X_1 \times X_2 \to Y: (x_1,x_2) \to F_1(x_1).F_2(x_2)$?

In fact it is not at all natural to have a property like this; the following simple counter-example proves it.

Let $F_1: C([0,1]) \to C([0,1]): x(t) \to e^{x(t)}$ and $F_2: C([0,1]) \to C([0,1]): y(t) \to e^{y(t)}$, then $F: C([0,1]) \times C([0,1]) \to C([0,1]): (x(t),y(t)) \to e^{x(t)+y(t)}$. Take n=1 and m=2.

The couple of abstract polynomials $(1+\frac{1}{3}x(t)\ ,\ 1-\frac{2}{3}x(t)+\frac{1}{6}x^2(t))$ belongs to the (1,2) abstract Padé approximant for F_1, $(1+\frac{1}{3}y(t)\ ,\ 1-\frac{2}{3}y(t)+\frac{1}{6}y^2(t))$ to the (1,2) abstract Padé approximant for F_2 and $(1+\frac{1}{3}(x+y)(t)\ ,\ 1-\frac{2}{3}(x+y)(t)+\frac{1}{6}(x+y)^2(t))$ to the (1,2) abstract Padé approximant for F. It is easy to see that

$[(1+\frac{1}{3}x(t)).(1+\frac{1}{3}y(t))\ ,\ (1-\frac{2}{3}x(t)+\frac{1}{6}x^2(t)).(1-\frac{2}{3}y(t)+\frac{1}{6}y^2(t))]$ does not belong to the (1,2) abstract Padé approximant for F.

Now let X be a Banach space and Y_i commutative Banach algebras. Consider nonlinear operators $F_i: X \to Y_i$, $i=1,\dots,q < \infty$ and $F: X \to \prod_{i=1}^{q} Y_i : x \to (F_i(x),\ i=1,\dots,q)$ where $\prod_{i=1}^{q} Y_i$ is a commutative Banach algebra with component-wise multiplication and normed by one of the Minkowski-norms $\|(y_1,\dots,y_q)\|_p$ $(1 \le p \le \infty)$.
We can obtain, by renorming, that $\|(I_1,\dots,I_q)\|_p = 1$ where I_i is the unit for the multiplication in Y_i.

Theorem I.12.2.:

Let $(\bigcap_{i=1}^{q} D(Q_i)) \cup (\bigcap_{i=1}^{q} D(P_i)) \neq \emptyset$ for the considered solution (P_i,Q_i) of (I.4.1) and (I.4.2) for F_i. Then $(P_{\star i},Q_{\star i})$ is the (n,m) abstract Padé approximant for F_i, $i=1,\dots,q$ if and only if

$$(P_\star,Q_\star) = \begin{bmatrix} P_{\star 1} & Q_{\star 1} \\ P_{\star 2} & Q_{\star 2} \\ \vdots & \vdots \\ P_{\star q} & Q_{\star q} \end{bmatrix}$$ is the (n,m) abstract Padé approximant for F.

Proof:

Since $(P_{\star i},\ Q_{\star i})$ is the (n,m) abstract Padé-approximant for F_i, abstract polynomials T_i exist such that $(P_i,Q_i) = (P_{\star i}.T_i,\ Q_{\star i}.T_i)$ satisfies (I.3.1) for F_i, in other words nonnegative constants K_i exist such that $\|[(F_i.Q_{\star i} - P_{\star i}).T_i](x)\| \le K_i\ \|x\|^{nm+n+m+1}$ in a neighbourhood of the origin, and this for $i = 1,\dots,q$.

Because $(\bigcap_{i=1}^{q} D(Q_i)) \cup (\bigcap_{i=1}^{q} D(P_i)) \neq \emptyset$ also $\bigcap_{i=1}^{q} D(T_i) \neq \emptyset$.

We use the Minkowski-norm $\|\ \|_p$ in $\prod_{i=1}^{q} Y_i$ for some p with $1 \le p \le \infty$.

Then for $p = 1$ let $K = \sum_i K_i$, for $p = \infty$ let $K = \max_i K_i$,

for $1 < p < \infty$ let $K = (\sum_i K_i^p)^{1/p}$ and we find

$$\| ([(F_i.Q_{\star i}-P_{\star i}).T_i](x),\ i = 1, \ldots, q) \|_p \le K.\|x\|^{nm+n+m+1}$$

in a neighbourhood of the origin.

Thus $(P, Q) = \begin{bmatrix} P_{\star 1}.T_1 & Q_{\star 1}.T_1 \\ \vdots & , \vdots \\ P_{\star q}.T_q & Q_{\star q}.T_q \end{bmatrix}$ satisfies (I.3.1) for

F and $D(P) \cup D(Q) \neq \emptyset$.

$\Leftarrow$

Since $(P_\star, Q_\star)$ is the (n,m) abstract Padé-approximant for F, an abstract polynomial T exists with $D(T) \neq \emptyset$ such that $[(F.Q_\star - P_\star).T](x) = O(x^{nm+n+m+1})$.

We write $(T)_i$ for the ith operator-component of T.

We know that $\| [(F_i.Q_{\star i}-P_{\star i}).(T)_i](x) \|_{(i)} \le \| [(F.Q_\star - P_\star).T](x) \|_p$

for $i = 1, \ldots, q$ and for whatever Minkowski-norm used in $\prod_{i=1}^{q} Y_i$.

So $(P_i, Q_i) = (P_{\star i}.(T)_i, Q_{\star i}.(T)_i)$ satisfies (I.3.1) for F_i and

$D(P_i) \cup D(Q_i) \neq \emptyset$ since $D(P_i) \supset \bigcap_{i=1}^{q} D(P_i)$ and $D(Q_i) \supset \bigcap_{i=1}^{q} D(Q_i)$. ■

Remark the fact that if $(\bigcap_{i=1}^{q} D(P_i)) \cup (\bigcap_{i=1}^{q} D(Q_i)) = \emptyset$, we cannot find x in X where the q solutions (P_i, Q_i) of (I.3.1) for F_i can be used simultaneously.

It is useless then to consider $(P, Q) = \begin{bmatrix} P_1 & Q_1 \\ \vdots & , \vdots \\ P_q & Q_q \end{bmatrix}$ since $D(P) \cup D(Q) = \emptyset$.

We illustrate the theorems I.12.1 and I.12.2 with an example.

Take $G : \mathbb{R}^2 \to \mathbb{R} : (x, y) \to \dfrac{xe^x - ye^y}{x - y}$.

The (1,1) APA for G is $\dfrac{x + y + 0.5(x^2 + 3\,xy + y^2)}{x + y - 0.5(x^2 + xy + y^2)}$.

For j=1: $x = 0$ and for j=2: $y = 0$

$G_1: \mathbb{R} \to \mathbb{R}: y \to e^y$ $\quad G_2: \mathbb{R} \to \mathbb{R}: x \to e^x$

Indeed the (1,1) APA for G_1 equals $\dfrac{1+\frac{1}{2}y}{1-\frac{1}{2}y}$ and for G_2 equals $\dfrac{1+\frac{1}{2}x}{1-\frac{1}{2}x}$.

Speaking again in terms of equivalence-classes, the couple of abstract polynomials $(x+y+0.5(x^2+3xy+y^2)\ ,\ x+y-0.5(x^2+xy+y^2))$ belongs to the (1,1) abstract Padé approximant for G. We already verified that $(1+10x-10.1y\ ,\ 1-10.1y)$ belonged to the (1,1) abstract Padé approximant for $F: \mathbb{R}^2 \to \mathbb{R}: (x,y) \to 1+\frac{1}{0.1-y}+\sin(xy)$.

Now
$$\begin{pmatrix} \dfrac{1+10x-10.1y}{x+y+0.5(x^2+3xy+y^2)} & , & \dfrac{1-10.1y}{x+y-0.5(x^2+xy+y^2)} \end{pmatrix}$$

belongs to the (1,1) abstract Padé-approximant for $\binom{F}{G}$.

We have to remark that the restrictive conditions formulated in all the theorems given in this chapter, are always fulfilled in the classical theory of Padé approximants for a univariate function.

CHAPTER II: MULTIVARIATE PADE APPROXIMANTS

§ 1. MOTIVATION

We will now study the multivariate Padé approximants ($X = \mathbb{R}^p$, $Y = \mathbb{R}$) more in detail. This is interesting because of several facts:

a) the irreducible form $\frac{1}{Q_\star}.P_\star$ is unique, Y contains no nilpotent elements and condition (I.10.1) is always satisfied, so all the theorems mentioned in chapter I are valid;

b) more similarities with the univariate Padé approximants can be proved, more properties can be formulated.

Besides those theoretical conclusions we will also compare our multivariate Padé approximant with other generalizations of the classical Padé approximant to multivariate functions, by means of many numerical examples.

Most of the times we will still use the notations

$$F(x) = \sum_{k=0}^{\infty} C_k x^k \text{ with } x \in \mathbb{R}^p$$

$$P(x) = \sum_{i=0}^{n} A_{nm+i} x^{nm+i}$$

$$Q(x) = \sum_{j=0}^{m} B_{nm+j} x^{nm+j}$$

$\frac{P_\star}{Q_\star}(x)$ for the irreducible form of $\frac{P}{Q}(x)$

where for the multivariate function F and the multilinear operators A_{nm+i} and B_{nm+j}:

$$C_k x^k = \sum_{k_1+\ldots+k_p=k} c_{k_1\ldots k_p} x_1^{k_1}\ldots x_p^{k_p}$$

$$\text{with } c_{k_1\ldots k_p} = \frac{1}{k_1!\ldots k_p!} \frac{\partial^k F(x_1,\ldots,x_p)}{\partial x_1^{k_1}\ldots\partial x_p^{k_p}}$$

$$A_{nm+i} x^{nm+i} = \sum_{i_1+\ldots+i_p = nm+i} a_{i_1\ldots i_p} x_1^{i_1}\ldots x_p^{i_p}$$

$$B_{nm+j} x^{nm+j} = \sum_{j_1+\ldots+j_p=nm+j} b_{j_1\ldots j_p} x_1^{j_1}\ldots x_p^{j_p}$$

§ 2. EXISTENCE OF A NONTRIVIAL SOLUTION

We already mentioned that the Padé-approximation problem (I.3.1) is equivalent with the solution of 2 linear systems of equations (I.4.1) and (I.4.2)

$$\begin{cases} C_o.B_{nm}\,x^{nm} = A_{nm}\,x^{nm} & \forall x \in \mathbb{R}^p \\ \vdots & \\ C_n\,x^n.B_{nm}\,x^{nm} + \ldots + C_o.B_{nm+n}\,x^{nm+n} = A_{nm+n}\,x^{nm+n} & \forall x \in \mathbb{R}^p \end{cases}$$

with $B_{nm+j}\,x^{nm+j} \equiv 0$ for $j > m$.

$$\begin{cases} C_{n+1}\,x^{n+1}.B_{nm}\,x^{nm} + \ldots + C_{n+1-m}\,x^{n+1-m}.B_{nm+m}\,x^{nm+m} = 0 & \forall x \in \mathbb{R}^p \\ \vdots & \\ C_{n+m}\,x^{n+m}.B_{nm}\,x^{nm} + \ldots + C_n\,x^n.B_{nm+m}\,x^{nm+m} = 0 & \forall x \in \mathbb{R}^p \end{cases}$$

with $C_k\,x^k \equiv 0$ for $k < o$.

Each term $B_j x^j = \sum_{j_1+\ldots+j_p=j} b_{j_1\ldots j_p}\, x_1^{j_1}\ldots x_p^{j_p}$ contains $\binom{p+j-1}{j}$ coefficients $b_{j_1\ldots j_p}$. So the homogeneous system (I.4.2) contains in total $N_u = \sum_{j=nm}^{nm+m} \binom{p+j-1}{j}$ unknown coefficients $b_{j_1\ldots j_p}$ of the B_{nm+j} $(j=o,\ldots,m)$.

The kth equation in (I.4.2) equates an (nm+n+k)-linear operator in p variables to zero. So it equates $\binom{p+nm+n+k-1}{nm+n+k}$ coefficients in that operator to zero. Thus in total we have $N_e = \sum_{k=1}^{n} \binom{p+nm+n+k-1}{nm+n+k}$ homogeneous equations.

It is easy to show that

$$N_e = \binom{p+nm+n+m}{nm+n+m} - \binom{p+nm+n}{nm+n}$$

and

$$N_u = \binom{p+nm+m}{nm+m} - \binom{p+nm-1}{nm-1}$$

if $nm > o$ and $N_u = \binom{p+m}{m}$ if $nm = o$.

a) For p=2: $N_u - N_e = 1$ and so one unknown can certainly be chosen and a nontrivial solution always exists.

b) If $p > 2$ the nontriviality of the solution is proved as follows.

Suppose that the matrix

$$\begin{pmatrix} C_{n+1}\, x^{n+1} & \cdots & C_{n+1-m}\, x^{n+1-m} \\ \vdots & & \\ C_{n+m}\, x^{n+m} & \cdots & C_n\, x^n \end{pmatrix}$$

of the homogeneous system (I.4.2) has rank k, in other words that a vector x exists in $\mathbb{R}^p$ such that the determinant of a kxk submatrix is nonzero. In any case $o \le k \le m$. The homogeneous system (I.4.2) can now be reduced to a homogeneous system of k equations in k+1 of the unknown $B_{nm+j}x^{nm+j}$ $(j=o,\ldots,m)$:

$$\left\{\begin{array}{l} \sum_{i=o}^{k} C_{n+h_1-j_i}\, x^{n+h_1-j_i}\, B_{nm+j_i}\, x^{nm+j_i} = 0 \\ \vdots \\ \sum_{i=o}^{k} C_{n+h_k-j_i}\, x^{n+h_k-j_i}\, B_{nm+j_i}\, x^{nm+j_i} = 0 \end{array}\right. \qquad \text{(II.2.1)}$$

with $1 \le h_i \le m$ for $i = 1, \ldots, k$

and $\begin{cases} o \le j_i \le m, & i = o, \ldots, k \\ j_o < \cdots < j_k \end{cases}$

In fact we have removed (m-k) rows and (m-k) columns in the coefficient matrix of system (I.4.2) to obtain the coefficient matrix of system (II.2.1). We will number the rows that we have removed $\bar{h}_1,\ldots,\bar{h}_{m-k}$ and the columns that we have removed $\bar{j}_1+1,\ldots,\bar{j}_{m-k}+1$ (notice that the rows that we have retained, are numbered $h_1,\ldots,h_k$ and the columns $j_o+1,\ldots,j_k+1$).

Write $\ell = n(m-k) + \sum_{i=1}^{m-k} (\bar{h}_i-\bar{j}_i)$. The determinant

$$\begin{vmatrix} C_{n+\bar{h}_1-\bar{j}_1}\, x^{n+\bar{h}_1-\bar{j}_1} & \cdots & C_{n+\bar{h}_1-\bar{j}_{m-k}}\, x^{n+\bar{h}_1-\bar{j}_{m-k}} \\ \vdots & & \\ C_{n+\bar{h}_{m-k}-\bar{j}_1}\, x^{n+\bar{h}_{m-k}-\bar{j}_1} & \cdots & C_{n+\bar{h}_{m-k}-\bar{j}_{m-k}}\, x^{n+\bar{h}_{m-k}-\bar{j}_{m-k}} \end{vmatrix}$$

is then a bounded ℓ-linear operator; it is easy to see that $o \le \ell \le nm+j_o$.

Let $E_\ell x^\ell$ be a nontrivial ℓ-linear bounded operator in $L((\mathbb{R}^p)^\ell,\mathbb{R})$. Then

$$B_{nm+j_0}\,x^{nm+j_0} = E_\ell\, x^\ell . \begin{vmatrix} C_{n+h_1-j_1}\,x^{n+h_1-j_1} & \cdots & C_{n+h_1-j_k}\,x^{n+h_1-j_k} \\ \vdots & & \\ C_{n+h_k-j_1}\,x^{n+h_k-j_1} & \cdots & C_{n+h_k-j_k}\,x^{n+h_k-j_k} \end{vmatrix}$$

and for $i = 1,\dots,k$

$$B_{nm+j_i}\,x^{nm+j_i} = E_\ell\, x^\ell . \begin{vmatrix} C_{n+h_1-j_1}\,x^{n+h_1-j_1} & \cdots & \boxed{-C_{n+h_1-j_0}\,x^{n+h_1-j_0}} & \cdots & C_{n+h_1-j_k}\,x^{n+h_1-j_k} \\ \vdots & & \vdots & & \vdots \\ C_{n+h_k-j_1}\,x^{n+h_k-j_1} & \cdots & \boxed{-C_{n+h_k-j_0}\,x^{n+h_k-j_0}} & \cdots & C_{n+h_k-j_k}\,x^{n+h_k-j_k} \end{vmatrix}$$

i^{th} column in $B_{nm+j_0}\,x^{nm+j_0}$
replaced by this column

is a nontrivial solution of (II.2.1) because one of the kxk determinants is nontrivial. If we choose the $B_{nm+\bar{j}_i}\,x^{nm+\bar{j}_i} = 0$ $(i=1,\dots,m-k)$ we have a nontrivial solution of the original homogeneous system (I.4.2).

§ 3. COVARIANCE PROPERTIES

Besides the properties mentioned in § 6. of chapter I, we can also prove the following theorems for multivariate Padé approximants.

Theorem II.3.1.:

Let $y_i = \dfrac{a_i x_i}{1+b_1x_1+\dots+b_px_p}$ for $i=1,\dots,p$ and let $y = (y_1,\dots,y_p)$.
Let $R_{n,n}$ for $F(x)$ be given by $\dfrac{P_\star}{Q_\star}(x)$ and let $G(x) := F(y)$, $R_\star(x) := P_\star(y)$, $S_\star(x) := Q_\star(y)$, then $R_{n,n}(x)$ for $G(x)$ is given by

$$\frac{[R_\star(x)(1+b_1\,x_1+\dots+b_p\,x_p)^k]}{[S_\star(x)(1+b_1\,x_1+\dots+b_p\,x_p)^k]} \quad \text{where } k = \max(\partial P_\star,\ \partial Q_\star).$$

Proof:

Because of theorem I.5.3, there exists a positive integer t_0, $n^2 - \partial_0 Q_\star \leq t_0 \leq n^2 - \partial_0 Q_\star + \min(n - \partial_1 P_\star, n - \partial_1 Q_\star)$ and a non-trivial symmetric t_0-linear bounded operator L_{t_0} such that $(P_\star . L_{t_0}, Q_\star . L_{t_0})$ satisfies (I.3.1) for the operator F.

We write $L_{t_0}(y) = \dfrac{L_{t_0}(a_1 x_1, \ldots, a_p x_p)}{(1 + b_1 x_1 + \ldots + b_p x_p)^{t_0}} = \dfrac{\overline{L}_{t_0}(x)}{(1 + b_1 x_1 + \ldots + b_p x_p)^{t_0}}$

Let $k = \max(\partial P_\star, \partial Q_\star)$.

Then $\partial_0(R_\star . \overline{L}_{t_0} . (1+b_1x_1+\ldots+b_px_p)^k) \geq \partial_0(P_\star . L_{t_0}) \geq n^2$

$\partial_0(S_\star . \overline{L}_{t_0} . (1+b_1x_1+\ldots+b_px_p)^k) \geq \partial_0(Q_\star . L_{t_0}) \geq n^2$

and $\max[\partial(R_\star . \overline{L}_{t_0} . (1+b_1x_1+\ldots+b_px_p)^k), \partial(S_\star . \overline{L}_{t_0} . (1+b_1x_1+\ldots+b_px_p)^k)]$

$$\leq k + t_0 \leq n^2 + n.$$

Also $(1+b_1x_1+\ldots+b_px_p)^k [(G.S_\star - R_\star).\overline{L}_{t_0}](x) =$

$$=[(F.Q_\star - P_\star).L_{t_0}](y).(1+b_1x_1+\ldots+b_px_p)^{k+t_0}$$

$$= O(y^{n^2+2n+1}).(1+b_1x_1+\ldots+b_px_p)^{k+t_0}$$

$$= O(x^{n^2+2n+1})$$

Thus $(R, S) = (R_\star . \overline{L}_{t_0} . (1+b_1x_1+\ldots+b_px_p)^k, S_\star . \overline{L}_{t_0} . (1+b_1x_1+\ldots+b_px_p)^k)$

satisfies (I.3.1) for the operator G.

We will now show that the irreducible form of $(\frac{1}{S}.R)(x)$ is

$$\frac{1}{[S_\star(x).(1+b_1x_1+\ldots+b_px_p)^k]} . [R_\star(x).(1+b_1x_1+\ldots+b_px_p)^k].$$

The factors $(1+b_1x_1+\ldots+b_px_p)^k$ are necessary because $R_\star(x)$ and $S_\star(x)$ are rational functions of the x_i, not polynomial.

Suppose $R_\star(x)(1+b_1x_1+\ldots+b_px_p)^k = U(x).V(x)$ and

$S_\star(x)(1+b_1x_1+\ldots+b_px_p)^k = U(x).W(x)$ with $\partial U \geq 1$.

Since $\dfrac{a_1x_1}{y_1} = \dfrac{a_2x_2}{y_2} = \ldots = \dfrac{a_px_p}{y_p} = 1 + \sum_{i=1}^{p} b_ix_i$ we know that

$$x_i = \frac{a_p . y_i}{a_i . y_p} . x_p \text{ for } i = 1,\dots,p.$$

Consequently $1 + \sum_{i=1}^{p} b_i x_i = 1 + x_p \sum_{i=1}^{p} b_i \frac{a_p y_i}{a_i y_p} = \frac{a_p x_p}{y_p}$

or $x_p = 1 \Big/ \left(\frac{a_p}{y_p} - b_p - \sum_{i=1}^{p-1} b_i \frac{a_p y_i}{y_p a_i}\right)$.

So we can write $\sum_{i=1}^{p} b_i\, x_i = \left(\sum_{i=1}^{p} b_i \frac{a_p y_i}{a_i y_p}\right) \Big/ \left(\frac{a_p}{y_p} - b_p - \sum_{i=1}^{p-1} b_i \frac{a_p y_i}{y_p a_i}\right)$

and $1 + \sum_{i=1}^{p} b_i x_i = 1 \Big/ \left(1 - \sum_{i=1}^{p} b_i \frac{y_i}{a_i}\right)$ and $x_i = \dfrac{y_i}{a_i\left(1 - \sum_{i=1}^{p} b_i \frac{y_i}{a_i}\right)}$.

Thus $R_\star(x) = P_\star(y)$ and $S_\star(x) = Q_\star(y)$ implies that

$P_\star(y) = U(x).V(x).\left(1 - \frac{b_1}{a_1} y_1 - \dots - \frac{b_p}{a_p} y_p\right)^k$ and

$Q_\star(y) = U(x).W(x).\left(1 - \frac{b_1}{a_1} y_1 - \dots - \frac{b_p}{a_p} y_p\right)$ and thus that

$P_\star(y) = \overline{U}(y).\overline{V}(y)$ and $Q_\star(y) = \overline{U}(y).\overline{W}(y)$ with

$$\begin{cases} \overline{U}(y) = U\left(\dfrac{y_1}{a_1\left(1-\frac{b_1}{a_1}y_1-\dots-\frac{b_p}{a_p}y_p\right)}, \dots, \dfrac{y_p}{a_p\left(1-\frac{b_1}{a_1}y_1-\dots-\frac{b_p}{a_p}y_p\right)}\right).\left(1-\frac{b_1}{a_1}y_1-\dots-\frac{b_p}{a_p}y_p\right)^{\overline{k}} \\ \overline{k} = \partial U \\ \overline{V}(y) = V(x_1,\dots,x_p).\left(1-\frac{b_1}{a_1}y_1-\dots-\frac{b_p}{a_p}y_p\right)^{k-\overline{k}} \qquad (\overline{k} + \partial V \le k) \\ \overline{W}(y) = W(x_1,\dots,x_p).\left(1-\frac{b_1}{a_1}y_1-\dots-\frac{b_p}{a_p}y_p\right)^{k-\overline{k}} \qquad (\overline{k} + \partial W \le k) \end{cases}$$

This contradicts the fact that $\frac{1}{Q_\star}.P_\star$ is irreducible.

Remark also that if $Q_\star(0) = 1$ then $S_\star(0) = 1$.

■

Theorem II.3.2.:

If $F(x) = \frac{G(x)}{H(x)}$ with $G(x) = \sum_{i=0}^{n} d_i x^i$ and $H(x) = \sum_{j=0}^{m} e_j x^j$ where $e_o \neq o$ and

where $d_i\ x^i = \sum_{i_1+\ldots+i_p=i} d_{i_1\ldots i_p}\ x_1^{i_1}\ldots x_p^{i_p}$

$e_j\ x^j = \sum_{j_1+\ldots+j_p=j} e_{j_1\ldots j_p}\ x_1^{j_1}\ldots x_p^{j_p}$

then for $F(x)$ irreducible we have $R_{n,m} = F(x)$.

Proof:

For $F(x) = \frac{G(x)}{H(x)}$ we can write $(F.H-G)(x) = O(x^{nm+n+m+1})$.

If $R_{n,m} = \frac{P_\star}{Q_\star}(x)$ then there exists a multivariate polynomial $T(x) \not\equiv 0$ such that $(P_\star.T, Q_\star.T)$ satisfies (I.3.1) for the rational funtion $F(x)$, in other words such that $(F.Q_\star.T-P_\star.T)(x) = O(x^{nm+n+m+1})$. Because of the equivalence-property of solutions of the Padé-approximation problem we can conclude that $(Q_\star.T.G)(x) = (P_\star.T.H)(x)$ for all x in $\mathbb{R}^p$, and hence that $(Q_\star.G)(x) = (P_\star.H)(x)$ for all x in $\mathbb{R}^p$. Using the unique factorisation of multivariate polynomials we can immediately write that $P_\star = G$ and $Q_\star = H$.

∎

This property will be referred to as the consistency property.

§ 4. NEAR-TOEPLITZ STRUCTURE OF THE HOMOGENEOUS SYSTEM

4.1. *Displacement rank*

For the sake of simplicity we restrict ourselves now to the case of two variables. To examine the special structure of the matrix of the homogeneous system, which we shall denote by H, we introduce the following notations:

for $Q(x,y) = \sum_{i+j=nm}^{nm+m} b_{ij}\ x^i\ y^j$ we write

$$B_{nm} = \begin{bmatrix} b_{nm,o} \\ b_{nm-1,1} \\ \vdots \\ b_{o,nm} \end{bmatrix} \quad B_{nm+1} = \begin{bmatrix} b_{nm+1,o} \\ b_{nm,1} \\ \vdots \\ b_{o,nm+1} \end{bmatrix} \cdots \quad B_{nm+m} = \begin{bmatrix} b_{nm+m,o} \\ b_{nm+m-1,1} \\ \vdots \\ b_{o,nm+m} \end{bmatrix}$$

When we write down the equations equivalent with condition (I.3.1), the set of homogeneous equations in the unknown b_{ij} is

$$H \begin{bmatrix} B_{nm} \\ \vdots \\ B_{nm+m} \end{bmatrix} = 0$$

with

$$H = \begin{bmatrix} H_{n+1,nm} & H_{n,nm+1} & \cdots & H_{n+1-m,nm+m} \\ H_{n+2,nm} & \cdots & & \\ \vdots & & & \vdots \\ H_{n+m,nm} & \cdots & & H_{n,nm+m} \end{bmatrix}$$

where $H_{i,j}$ is a matrix with $(i+j+1)$ rows and $(j+1)$ columns and the first column equal to the transpose of $(c_{i,o}\ c_{i-1,1}\ \ldots\ c_{1,i-1}\ c_{o,i}\ o \ldots o)$ and the next columns equal to their previous column but with all the elements shifted down one place and a zero added on top. The matrix H has $N_e = \binom{nm+n+m+2}{2} - \binom{nm+n+2}{2}$ rows and one more columns than rows. To calculate the displacement rank $\alpha(H)$ of H, we have to construct the lower shifted difference

$$H - \overline{H} = \begin{bmatrix} h_{1,1} & \cdots & h_{1,N_e+1} \\ \vdots & & \\ & & \\ h_{N_e,1} & \cdots & h_{N_e,N_e+1} \end{bmatrix} - \begin{bmatrix} o & \cdots & & o \\ & h_{1,1} & \cdots & h_{1,N_e} \\ & \vdots & & \vdots \\ o & h_{N_e-1,1} & \cdots & h_{N_e-1,N_e} \end{bmatrix} = \begin{bmatrix} h_{1,1} & \cdots & h_{1,N_e+1} \\ \vdots & & \\ & & \delta H \\ h_{N_e,1} & & \end{bmatrix}$$

Now $\alpha(H) = \text{rank}(\delta H) + 2$ [31]. The concept of displacement rank serves as a measure of how close to toeplitz a given matrix is, since $\text{rank}(\delta H) = o$ if H is actually a toeplitz matrix.

Theorem II.4.1.:

The displacement rank of the matrix H is at most m+2.

Proof:

Let us write down the matrix H more explicitly:

$$
H = \left(\begin{array}{ccc|ccc|c|ccc}
c_{n+1,0} & 0\;\cdots & 0 & c_{n,0} & 0\;\cdots & 0 & & c_{n+1-m,0} & 0\;\cdots & 0 \\
\vdots & \ddots & \vdots & \vdots & & \vdots & \cdots & \vdots & & \vdots \\
c_{0,n+1} & & 0 & c_{0,n} & & 0 & & c_{0,n+1-m} & & 0 \\
0 & & c_{n+1,0} & 0 & & c_{n,0} & & 0 & & c_{n+1-m,0} \\
\vdots & \ddots & \vdots & \vdots & & \vdots & & \vdots & & \vdots \\
0\;\cdots & 0 & c_{0,n+1} & 0 & \cdots\;0 & c_{0,n} & & 0\;\cdots & 0 & c_{0,n+1-m} \\
\hline
 & \vdots & & & & & & & & \\
\hline
c_{n+m,0} & 0\;\cdots & 0 & & & & & & & \\
\vdots & & \vdots & & & & & & & \\
c_{0,n+m} & & 0 & & \cdots & & & & \cdots & \\
0 & & c_{n+m,0} & & & & & & & \\
\vdots & & \vdots & & & & & & & \\
0\;\cdots & 0 & c_{0,n+m} & & & & & & &
\end{array}\right)
$$

Then δH has the following structure:

$\delta H = (\Delta_1\ \Delta_2\ \cdots\ \Delta_{m+1})$, where

Δ_1 has $(N_e - 1)$ rows and nm columns,

Δ_i has $(N_e - 1)$ rows and (nm+i) columns for $i = 2,\ldots,m+1$

and only the first column in Δ_i with $i \geq 2$ contains nonzero elements; all the other elements in δH equal zero.

So rank $(\delta H) \leq m$ and this proves our theorem.

∎

It is easy to see that in the multivariate case the coefficient matrix of the homogeneous system is also a matrix with low displacement rank. Consequently algorithms can be used where the solution of the linear system in the Padé-approximation problem is given in less operations than usual [19] , i.e. in $O(\alpha(H)N_e^2)$ operations instead of $O(N_e^3)$ operations.

4.2. *Numerical examples*

We will illustrate the preceding theorems with some simple examples.

Consider $F(x,y) = 1 + \frac{x}{0.1-y} + \sin(xy)$

a) The (1,1) APA is $\frac{1+10x-10.1y}{1-10.1y}$ with

$$H = \begin{bmatrix} 0 & 0 & 10 & 0 & 0 \\ 101 & 0 & 0 & 10 & 0 \\ 0 & 101 & 0 & 0 & 10 \\ 0 & 0 & 0 & 0 & 0 \end{bmatrix} \quad \text{and } \alpha(H) = 3.$$

b) The (4,2) APA is $\frac{1+10x-10y+xy-10xy^2}{1-10y}$ with

$$H = \begin{bmatrix} H_{5,8} & H_{4,9} & H_{3,10} \\ H_{6,8} & H_{5,9} & H_{4,10} \end{bmatrix} \quad \text{and } \alpha(H) = 4,$$

where $H_{5,8} = 10^5 \, (\delta_{i,j+4})$ a 14 x 9 matrix

$H_{4,9} = 10^4 \, (\delta_{i,j+3})$ a 14 x 10 matrix

$H_{3,10} = 10^3 \, (\delta_{i,j+2})$ a 14 x 11 matrix

$H_{6,8} = 10^6 \, (\delta_{i,j+5}) - \frac{1}{6}(\delta_{i,j+3})$ a 15 x 9 matrix

$H_{5,9} = 10^5 \, (\delta_{i,j+4})$ a 15 x 10 matrix

$H_{4,10} = 10^4 \, (\delta_{i,j+3})$ a 15 x 11 matrix

and $\delta_{i,j}$ is the Kronecker symbol (here used in rectangular matrices).

§ 5. THREE-TERM IDENTITIES

5.1. *Cross ratios*

A cross ratio is a ratio of the form

$$\frac{(r_1-r_2)(r_3-r_4)}{(r_1-r_3)(r_2-r_4)} = R \tag{II.5.1}$$

Each of the four r_i appears in the numerator as well as in the denominator.
If we compute (II.5.1) when the r_i are the values of four adjacent entries in the Padé table for a given value $\bar{x} = (\bar{x}_1,\ldots,\bar{x}_p)$, then we can use the two-term identities (I.7.1)-(I.7.4) to simplify R. Consider for instance the four entries in the Padé table given in figure II.5.1

$\frac{P_{[n,m]}}{Q_{[n,m]}}(\bar{x}) = r_1$	$\frac{P_{[n,m+1]}}{Q_{[n,m+1]}}(\bar{x}) = r_2$
$\frac{P_{[n+1,m]}}{Q_{[n+1,m]}}(\bar{x}) = r_3$	$\frac{P_{[n+1,m+1]}}{Q_{[n+1,m+1]}}(\bar{x}) = r_4$

Figure II.5.1.

Then

$$R = \frac{H_{m+1}(C_n\bar{x}^n).H_{m+1}(C_{n+2}\bar{x}^{n+2})}{H_m(C_{n+1}\bar{x}^{n+1}).H_{m+2}(C_{n+1}\bar{x}^{n+1})}$$

For the entries given in the figures II.5.2 and II.5.3 we find respectively

$$R = \frac{H_m(C_{n+1}\bar{x}^{n+1}).H_{m+1}(C_{n+1}\bar{x}^{n+1})}{H_{m+1}(C_n\bar{x}^n).H_m(C_{n+2}\bar{x}^{n+2})}$$

and

$$R = \frac{H_{m+1}(C_n\bar{x}^n).H_{m+1}(C_{n+1}\bar{x}^{n+1})}{H_m(C_{n+1}\bar{x}^{n+1}).H_{m+2}(C_n\bar{x}^n)}$$

	$\frac{P_{[n,m]}}{Q_{[n,m]}}(\bar{x}) = r_3$	$\frac{P_{[n,m+1]}}{Q_{[n,m+1]}}(\bar{x}) = r_4$
$\frac{P_{[n+1,m-1]}}{Q_{[n+1,m-1]}}(\bar{x}) = r_1$	$\frac{P_{[n+1,m]}}{Q_{[n+1,m]}}(\bar{x}) = r_2$	

Figure II.5.2.

	$\frac{P_{[n-1,m+1]}}{Q_{[n-1,m+1]}}(\bar{x}) = r_1$
$\frac{P_{[n,m]}}{Q_{[n,m]}}(\bar{x}) = r_3$	$\frac{P_{[n,m+1]}}{Q_{[n,m+1]}}(\bar{x}) = r_2$
$\frac{P_{[n+1,m]}}{Q_{[n+1,m]}}(\bar{x}) = r_4$	

Figure II.5.3.

Many more cross ratios can be calculated by means of the given two-term identities, but we give these examples because we shall use them now to derive some three-term identities.

2. *Three-term identities*

The cross ratio (II.5.1) can be solved for one of the r_i, say r_4, in terms of the other three r_i and R. We get

$$r_4 = \frac{r_3(r_2-r_1) - R\, r_2(r_3-r_1)}{(r_2-r_1) - R\,(r_3-r_1)}$$

If we use again figure II.5.1 we find

$$\frac{P_{[n+1,m+1]}}{Q_{[n+1,m+1]}}(\bar{x}) = \frac{P_{[n+1,m]}(\bar{x}).H_{m+2}(C_{n+1}\bar{x}^{n+1}) - P_{[n,m+1]}(\bar{x}).H_{m+1}(C_{n+2}\bar{x}^{n+2})}{Q_{[n+1,m]}(\bar{x}).H_{m+2}(C_{n+1}\bar{x}^{n+1}) - Q_{[n,m+1]}(\bar{x}).H_{m+1}(C_{n+2}\bar{x}^{n+2})}$$

So $\frac{P_{[n+1,m+1]}}{Q_{[n+1,m+1]}}(\bar{x})$ can be calculated by means of $P_{[n+1,m]}(\bar{x})$, $Q_{[n+1,m]}(\bar{x})$, $P_{[n,m+1]}(\bar{x})$ and $Q_{[n,m+1]}(\bar{x})$; we shall indicate this by

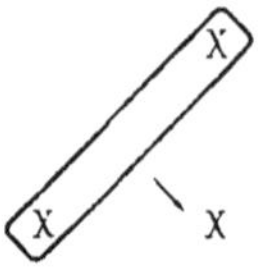

For the figures II.5.2 and II.5.3 we get respectively

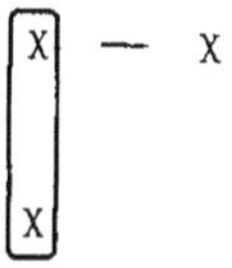

and

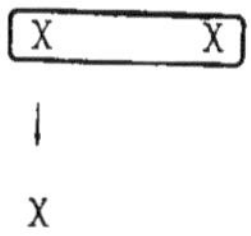

When we would calculate the cross ratios for the figures II.5.4, II.5.5 and II.5.6 we would find respectively

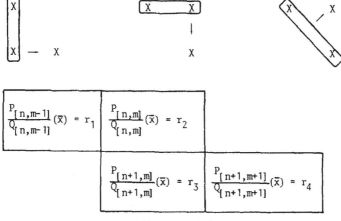

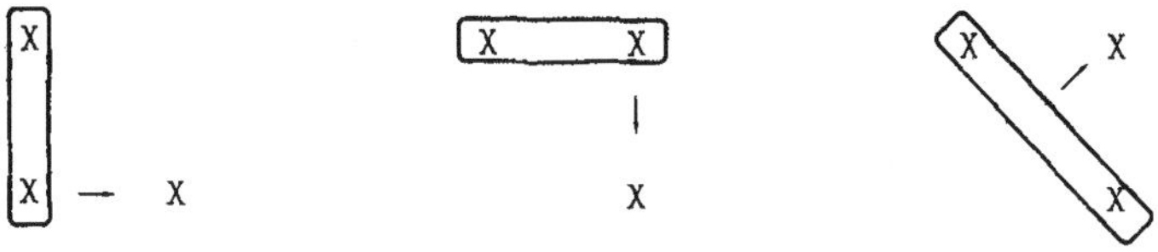

$\frac{P_{[n,m-1]}}{Q_{[n,m-1]}}(\bar{x}) = r_1$	$\frac{P_{[n,m]}}{Q_{[n,m]}}(\bar{x}) = r_2$	
	$\frac{P_{[n+1,m]}}{Q_{[n+1,m]}}(\bar{x}) = r_3$	$\frac{P_{[n+1,m+1]}}{Q_{[n+1,m+1]}}(\bar{x}) = r_4$

Figure II.5.4.

$\frac{P_{[n-1,m]}}{Q_{[n-1,m]}}(\bar{x}) = r_1$	
$\frac{P_{[n,m]}}{Q_{[n,m]}}(\bar{x}) = r_2$	$\frac{P_{[n,m+1]}}{Q_{[n,m+1]}}(\bar{x}) = r_3$
	$\frac{P_{[n+1,m+1]}}{Q_{[n+1,m+1]}}(\bar{x}) = r_4$

Figure II.5.5.

$\frac{P_{[n,m]}}{Q_{[n,m]}}(\bar{x}) = r_3$	$\frac{P_{[n,m+1]}}{Q_{[n,m+1]}}(\bar{x}) = r_4$
$\frac{P_{[n+1,m]}}{Q_{[n+1,m]}}(\bar{x}) = r_1$	$\frac{P_{[n+1,m+1]}}{Q_{[n+1,m+1]}}(\bar{x}) = r_2$

Figure II.5.6.

Of course much more figures can be considered than the ones indicated here, but these few examples give an idea about the possibilities that exist.

§ 6. ACCELERATING THE CONVERGENCE OF A TABLE WITH MULTIPLE ENTRY

6.1. *Table with double entry*

The ε-algorithm has frequently been used to accelerate the convergence of a sequence $(T_i)_{i=0}^{\infty}$ in ℝ [49], which can in fact be considered as a table with single entry: construct the univariate function

$$F(x) = \sum_{i=0}^{\infty} c_i x^i$$

where

$$c_i = T_i - T_{i-1} \qquad (T_i = 0 \text{ for } i < 0)$$

and calculate the classical Padé approximants for F.
Since

$$F(1) = \lim_{i\to\infty} T_i$$

one evaluates these Padé approximants at $x = 1$.
Let us now first consider a table $(T_{ij})_{i,j=0}^{\infty}$ with double entry. To accelerate the convergence of $(T_{ij})_{i,j=0}^{\infty}$ to $\lim_{i,j\to\infty} T_{ij}$ we introduce

$$F(x,y) = \sum_{i,j=0}^{\infty} c_{ij} x^i y^j$$

with

$$c_{ij} = T_{ij} - T_{i,j-1} - T_{i-1,j} + T_{i-1,j-1} \qquad (T_{ij} = 0 \text{ for } i < 0 \text{ or } j < 0)$$

Clearly

$$F(1,1) = \lim_{i,j\to\infty} T_{ij}$$

Using the generalization of the ε-algorithm given in § 7. of chapter I, we can calculate multivariate Padé approximants for $F(x,y)$ and evaluate them at $(x,y) = (1,1)$.
If we denote by

$$\mathcal{T}_n = \sum_{i+j=n} T_{ij} = T_{n,0} + T_{n-1,1} + \ldots + T_{1,n-1} + T_{0,n}$$

then the partial sums

$$F_n(1,1) = \sum_{i+j=0}^{n} c_{ij} = \mathcal{T}_n - \mathcal{T}_{n-1} \qquad n=0,\ldots,\infty$$

are the $\varepsilon_0^{(n)}$ to start the ε-algorithm with.
An application to accelerate the convergence in quadrature problems will be given at the end of this paragraph, but first we will generalize the idea for a table with multiple entry.

6.2. *Table with multiple entry*

Let us denote by $(T_{i_1 \ldots i_p})_{i_1,\ldots,i_p=0}^{\infty}$ a table with multiple entry.
We define

$$F(x_1,\dots,x_p) = \sum_{i_1,\dots,i_p=0}^{\infty} c_{i_1\dots i_p} x_1^{i_1}\dots x_p^{i_p}$$

with

$$c_{i_1\dots i_p} = T_{i_1\dots i_p} - \sum_{j=1}^{p} T_{i_1\dots(i_j-1)\dots i_p}$$

$$+ \sum_{\substack{j,k=1\\ j\neq k}}^{p} T_{i_1\dots i_{j-1}(i_j-1)i_{j+1}\dots i_{k-1}(i_k-1)i_{k+1}\dots i_p}$$

$$- \dots + (-1)^p T_{(i_1-1)\dots(i_p-1)}$$

It is easy to prove that

$$F(1,\dots,1) = \lim_{i_1,\dots,i_p\to\infty} T_{i_1\dots i_p}$$

Again multivariate Padé approximants for $F(x_1,\dots,x_p)$ can be calculated and evaluated at $(x_1,\dots,x_p) = (1,\dots,1)$ via the ε-algorithm.

Since

$$\sum_{i_1+\dots+i_p=n} c_{i_1\dots i_p} = \sum_{j=0}^{p} (-1)^j \binom{p}{j} T_{n-j}$$

where

$$T_n = \sum_{i_1+\dots+i_p=n} T_{i_1\dots i_p}$$

the $\varepsilon_0^{(n)}$ are now given by

$$F_n(1,\dots,1) = \sum_{i_1+\dots+i_p=0}^{n} c_{i_1\dots i_p} = \sum_{j=0}^{p-1} (-1)^j \binom{p-1}{j} T_{n-j}$$

6.3. *Applications*

Suppose one wants to calculate the integral of a function $F(x_1,\dots,x_p)$ on a bounded closed domain Ω of $\mathbb{R}^p$. Let $\Omega = [0,1] \times \dots \times [0,1] \subset \mathbb{R}^p$ for the sake of simplicity. The table $(T_{i_1\dots i_p})_{i_1,\dots,i_p=0}^{\infty}$ can be obtained for instance by subdividing the interval $[0,1]$ in the jth direction $(j=1,\dots,p)$ into 2^{i_j} intervals of equal length $h_j = \frac{1}{2^{i_j}}$ $(i_j=0,1,\dots)$.

Using the midpoint-rule one can then substitute approximations

$$\int_0^{h_1}\dots\int_0^{h_p} F(x_1,\dots,x_p)\, dx_1\dots dx_p = h_1h_2\dots h_p\, F(\frac{h_1}{2}, \frac{h_2}{2}, \dots, \frac{h_p}{2})$$

to calculate the $T_{i_1\dots i_p}$.

The column $\varepsilon_0^{(n)}$ $(n=0,1,2,\dots)$ in the ε-table given by

$$\varepsilon_0^{(n)} = \sum_{j=0}^{p-1} (-1)^j \binom{p-1}{j} T_{n-j}$$

was also used by Genz [22] to start the ε-algorithm for the approximate calculation of multidimensional integrals by means of extrapolation methods. He preferred this method to six other methods because of its simplicity and general use of fewer integrand evaluations [22]. But when he was using it he did regret that there was no link for the multidimensional problems with Padé approximants as there is in one dimension. Genz remarked that the construction and theory of the multivariate generalization of Padé approximants had only recently been developed by Chisholm and his staff, but that the Canterbury approximants were not particularly suitable for the problem of the extrapolation of sequences of approximations to multiple integrals. This paragraph has now put things together: the $\varepsilon_0^{(n)}$ are the partial sums of the multivariate function

$$F(x_1,\ldots,x_p) = \sum_{i_1,\ldots,i_p=0}^{\infty} c_{i_1\ldots i_p} x_1^{i_1}\ldots x_p^{i_p}$$

with the $c_{i_1\ldots i_p}$ defined above, and the $\varepsilon_{2m}^{(n-m)}$ are the (n,m) APA for that multivariate function, all evaluated in $(x_1,\ldots,x_p) = (1,\ldots,1)$. We will illustrate everything with some numerical results.

Let us now take p=2, $h_1=2^{-i}$, $h_2=2^{-j}$. Then

$$T_{ij} = \frac{1}{2^{i+j}} \left(\sum_{k=1}^{2^i} \sum_{\ell=1}^{2^j} F\left(\frac{2k-1}{2^{i+1}}, \frac{2\ell-1}{2^{j+1}}\right)\right)$$

For the first example we are going to consider, we have

$$F(x,y) = (x+y)^2$$

$$\int_0^1 \int_0^1 F(x,y)\, dx\, dy = \frac{7}{6} = 1.166666666666\ldots$$

$$T_{00} = 1$$

In table II.6.1 one can find some T_{ij} and some values of the (n,m) APA in (x,y)=(1,1). For the calculation of the (n,m) APA we need T_j, j=o,...,n+m. It is easy to see that the convergence is indeed improved.

$T_{10} = \frac{17}{16} = 1.0625$	(0,1) APA $= \frac{8}{7} = 1.142857142857\ldots$
$T_{11} = \frac{9}{8} = 1.125$	(1,1) APA $= \frac{7}{6} = 1.166666666666\ldots$
$T_{21} = \frac{73}{64} = 1.140625$	(2,1) APA $= \frac{7}{6} = 1.166666666666\ldots$
$T_{22} = \frac{37}{32} = 1.15625$	(3,1) APA $= \frac{7}{6} = 1.166666666666\ldots$

Table II.6.1.

What's more: substituting the explicit formula for the T_{ij} in the calculation of $\varepsilon_0^{(n)}$ one can easily check, using the expressions

$$\sum_{k=1}^{2^i} k^2 = \frac{1}{3} 2^{i-1} (2^i+1)(2^{i+1}+1) \text{ and } \sum_{k=1}^{2^i} k = 2^{i-1}(2^{i+1}+1)$$

that $\varepsilon_0^{(n)} = \frac{7}{6} - \frac{1}{6}(\frac{1}{4})^n$ for $n \geq 0$ which implies [9 pp. 45] that the value of the $(n+1,1)$ APA $= \varepsilon_2^{(n)} = \frac{7}{6}$ for $n \geq 0$.

As a second example we will approximate

$$\int_0^1 \int_0^1 \frac{1}{x+y} \, dx \, dy = 2\ell n 2 = 1.386294361119891$$

In table II.6.2 one can again find the T_{ij}, slowly converging to the exact value of the integral because of the singularity of the integrand in (0,0). The function-values of the (n,m) APA converge much faster.

T_{11} = 1.166666666667	(1,1) APA = 1.330294906166
T_{21} = 1.209102009102	(2,1) APA = 1.361763927710
T_{22} = 1.269047619048	(2,2) APA = 1.396395820203
T_{23} = 1.292977663088	(2,3) APA = 1.386056820469
T_{33} = 1.325743700744	(3,3) APA = 1.386872037696
T_{34} = 1.338426108120	(3,4) APA = 1.386481238969
T_{44} = 1.355532404415	(4,4) APA = 1.386308917778
T_{54} = 1.362055745711	(5,4) APA = 1.386298323641

Table II.6.2.

§ 7. COMPARISON WITH SOME OTHER TYPES OF MULTIVARIATE PADE APPROXIMANTS

We will restrict ourselves to the case of two variables because the generalization to more than two variables is straightforward. Many definitions exist that try to generalize the concept of Padé approximant to multivariate functions. However, the calculation of each type of multivariate Padé approximant $P_\star(x,y) / Q_\star(x,y)$ is based on:

$$(F.Q-P)(x,y) = \sum_{i,j=0}^{\infty} d_{ij} x^i y^j \text{ with } d_{ij} = 0 \text{ for } (i,j) \in E \subset \mathbb{N}^2.$$

We call E the interpolationset; the choice of E, P(x,y) and Q(x,y) determines the type of approximant.

If one wants the multivariate Padé approximant to satisfy the covariance properties I.6.1 and I.6.2, E must satisfy the inclusion property, i.e. if $(i,j) \in E$ then $([0,i] \times [0,j]) \cap \mathbb{N}^2 \subset E$.

We shall now briefly repeat the definition of some types of approximants and compare them theoretically and numerically with our abstract Padé approximants in the case $X = \mathbb{R}^p$ and $Y = \mathbb{R}$.

7.1. *General order Padé-type rational approximants introduced by Levin* [34]

We briefly repeat some notations and definitions given by Levin.

Given a subset D of $\mathbb{Z}^2$ we define:

a) the complement $\overline{D} = \mathbb{Z}^2 \setminus D$

b) the (i,j)-translation of D as $D_{ij} = \{(k,n) \mid (k+i,n+j) \in D\}$

c) the non-negative part of D as $D^+ = D \cap \mathbb{N}^2$

To any subset D such that D^+ is a finite set we associate polynomials

$$\sum_{(i,j)\in D^+} b_{ij} x^i y^j$$

We call D the rank of the polynomials.

Given the double power series

$$F(x,y) = \sum_{i,j=0}^{\infty} c_{ij} x^i y^j$$

we will choose three subsets N, D and E of $\mathbb{Z}^2$ and construct an $[N/D]_E$ approximation to F(x,y) as follows

$$P(x,y) = \sum_{(i,j)\in N^+} a_{ij} x^i y^j \quad \text{(N from "numerator")}$$

$$Q(x,y) = \sum_{(i,j)\in D^+} b_{ij} x^i y^j \quad \text{(D from "denominator")}$$

$$(F.Q-P)(x,y) = \sum_{(i,j)\in E^+} d_{ij} x^i y^j \quad \text{(E from "equations")} \tag{II.7.1}$$

We select N, D and E such that

a) $D \subset \mathbb{N}^2$ has m elements, numbered $(i_1,j_1),\ldots,(i_m,j_m)$

b) $N \subset E$ and $H = E \setminus N$ has m-1 elements in $\mathbb{N}^2$, numbered $(h_2,k_2),\ldots,(h_m,k_m)$
   (H from "homogeneous equations")

Then P(x,y) and Q(x,y) defined by equations (II.7.1), are given by

$$P(x,y) = \begin{vmatrix} x^{i_1} y^{j_1} N_{i_1j_1}(x,y) & x^{i_2} y^{j_2} N_{i_2j_2}(x,y) & \ldots & x^{i_m} y^{j_m} N_{i_mj_m}(x,y) \\ c_{h_2-i_1,k_2-j_1} & c_{h_2-i_2,k_2-j_2} & \ldots & c_{h_2-i_m,k_2-j_m} \\ c_{h_3-i_1,k_3-j_1} & c_{h_3-i_2,k_3-j_2} & \ldots & c_{h_3-i_m,k_3-j_m} \\ \vdots & \vdots & & \vdots \\ c_{h_m-i_1,k_m-j_1} & c_{h_m-i_2,k_m-j_2} & \ldots & c_{h_m-i_m,k_m-j_m} \end{vmatrix}$$

where $N_{i_\ell j_\ell}(x,y) = \sum_{(i,j)\in N^+_{i_\ell j_\ell}} c_{ij} x^i y^j \ (\ell = 1,\ldots,m)$ and

$$Q(x,y) = \begin{vmatrix} x^{i_1} y^{j_1} & x^{i_2} y^{j_2} & \dots & x^{i_m} y^{j_m} \\ c_{h_2-i_1,k_2-j_1} & c_{h_2-i_2,k_2-j_2} & \dots & c_{h_2-i_m,k_2-j_m} \\ \vdots & \vdots & & \vdots \\ c_{h_m-i_1,k_m-j_1} & c_{h_m-i_2,k_m-j_2} & \dots & c_{h_m-i_m,k_m-j_m} \end{vmatrix}$$

When we make the following choices for the sets N, D and E:

$$N = \{(i,j) \mid i,j \in \mathbb{N},\ nm \le i+j \le nm+n\}$$

$$D = \{(i,j) \mid i,j \in \mathbb{N},\ nm \le i+j \le nm+m\}$$

$$E = \{(i,j) \mid i,j \in \mathbb{N},\ nm \le i+j \le nm+n+m\}$$

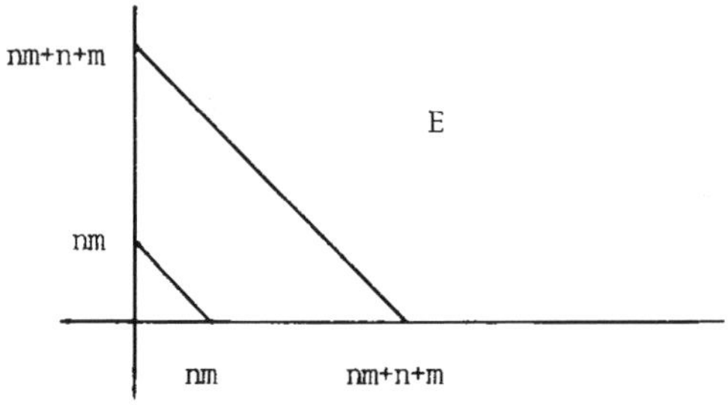

we get precisely the (n,m) abstract Padé approximant; the set $H = E \setminus N$ has one element less than the set D, as required.

7.2. *Canterbury approximants, Lutterodt approximants and Karlsson-Wallin approximants*

We are going to compare abstract Padé approximants (APA) for F(x,y) with Chisholm diagonal [13] approximants (CA), Hughes Jones off-diagonal [29,30] approximants (HJA), Lutterodt [37] approximants (LA), Lutterodt approximants of type B^1 [36] (LAB^1) and Karlsson-Wallin [32] approximants (KWA).

For the Canterbury approximants (i.e. CA and HJA) we have

$$P(x,y) = \sum_{i=0}^{n_1} \sum_{j=0}^{n_2} a_{ij} x^i y^j$$

$$Q(x,y) = \sum_{i=0}^{m_1} \sum_{j=0}^{m_2} b_{ij} x^i y^j$$

$$E = \{(i,j) \mid o \le i \le \max(n_1,m_1),\ o \le j \le \min(n_2,m_2)\}$$
$$\cup \{(i,j) \mid o \le i \le \min(n_1,m_1),\ o \le j \le \max(n_2,m_2)\}$$
$$\cup \{(i,j) \mid \max(n_2,m_2) < j \le n_2+m_2,\ \max(n_2,m_2) < i+j \le n_2+m_2,\ o \le i \le \min(n_1,m_1)\}$$
$$\cup \{(i,j) \mid \max(n_1,m_1) < i \le n_1+m_1,\ \max(n_1,m_1) < i+j \le n_1+m_1,\ o \le j \le \min(n_2,m_2)\}$$

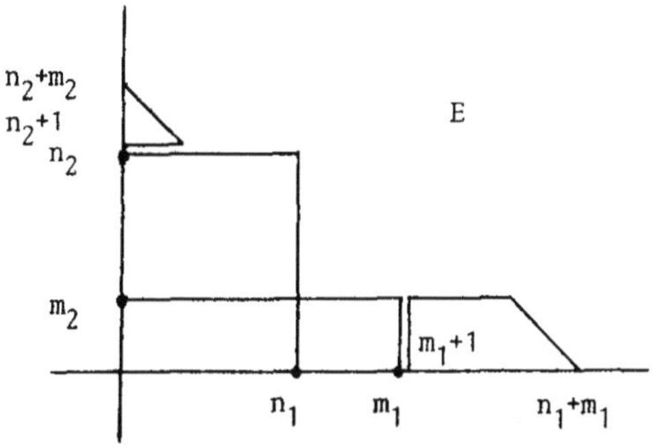

$$d_{ij} = o \text{ for } (i,j) \in E$$

$$d_{n_1+m_1+1-\ell,\ell} + d_{\ell,n_2+m_2+1-\ell} = o \text{ for } \ell = 1,\dots,\min(n_1,m_1,n_2,m_2)$$

For the Lutterodt approximants we have

$$P(x,y) = \sum_{i=o}^{n_1} \sum_{j=o}^{n_2} a_{ij}\, x^i\, y^j$$

$$Q(x,y) = \sum_{i=o}^{m_1} \sum_{j=o}^{m_2} b_{ij}\, x^i\, y^j$$

$$E \supset [0,n_1] \times [0,n_2] \cap \mathbb{N}^2$$

E satisfies the inclusion-property and contains exactly

$(n_1+1)(n_2+1) + (m_1+1)(m_2+1) - 1$ elements

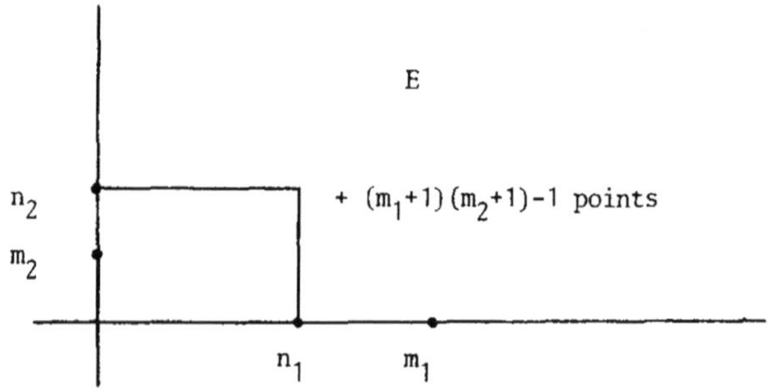

and for the Lutterodt approximants of type B^1

$$E = \{(i,j) \mid 0 \le i \le n_1,\ 0 \le j \le n_2\}$$
$$\cup \{(i,j) \mid n_1+1 \le i \le n_1+m_1,\ n_2+1 \le j \le n_2+m_2\}$$
$$\cup \{(i,0) \mid n_1+1 \le i \le n_1+m_1\}$$
$$\cup \{(0,j) \mid n_2+1 \le j \le n_2+m_2\}$$

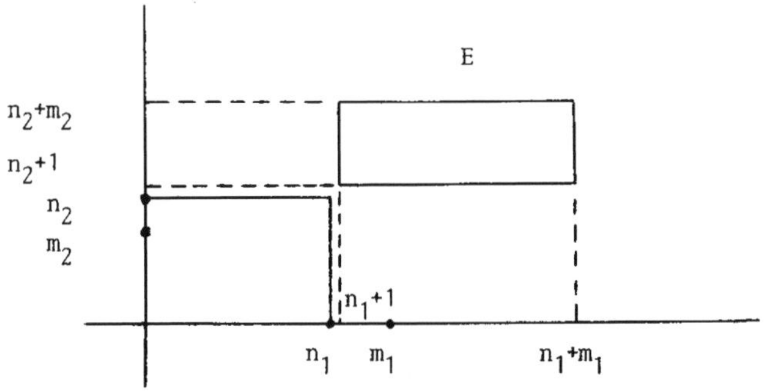

For the Karlsson-Wallin approximants we have

$$P(x,y) = \sum_{i+j=0}^{n} a_{ij}\, x^i\, y^j$$

$$Q(x,y) = \sum_{i+j=0}^{m} b_{ij}\, x^i\, y^j$$

$$E \supset \{(i,j) \mid i+j \le n\}$$

E satisfies the inclusion-property and contains at least

$\frac{1}{2}(n+1)(n+2) + \frac{1}{2}(m+1)(m+2) - 1$ elements

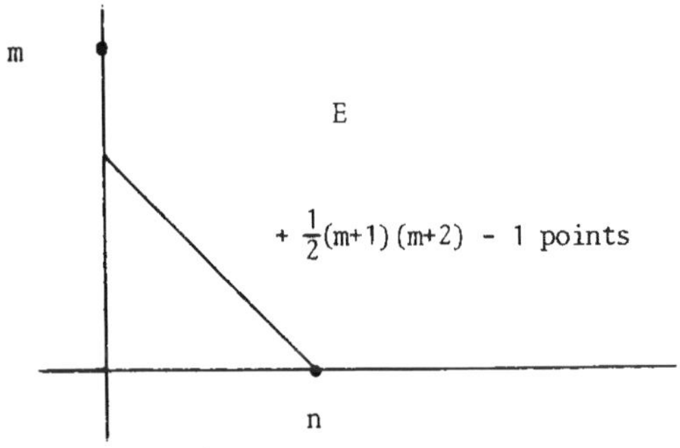

The following scheme summarizes the properties satisfied by each type of approximant

	CA and HJA	LA	LAB[1]	KWA	APA
Unicity	Under certain conditions on the c_{ij}	With respect to a given E only if for a chosen normalization the homogeneous system has a unique solution	Same remark as for LA	If E contains as many points as possible from some given enumeration of the points in $\mathbb{N} \times \mathbb{N}$	Yes
Property I.6.1	Yes	Yes	No	Yes	Yes
Property I.6.2	For CA	For $(n_1,n_2)/(n_1,n_2)$	No	Yes	Yes
Projection property	Yes	Only if $E \supset \{(o,j) \mid o \le j \le n_2+m_2\} \cup \{(i,o) \mid o \le i \le n_1+m_1\}$	Yes	Only if $E \supset \{(o,j) \mid o \le j \le n+m\} \cup \{(i,o) \mid o \le i \le n+m\}$	Yes
Symmetry	For $(n,n)/(m,m)$	For $(n,n)/(m,m)$ if (II.7.2) is satisfied	For $(n,n)/(m,m)$	If (II.7.2) is satisfied	Yes
Variable changes	$\bar{x} = \frac{ax}{1+b_1x}$ $\bar{y} = \frac{ay}{1+b_2y}$ for CA	$\bar{x} = \frac{a_1x}{1+b_1x}$ $\bar{y} = \frac{a_2y}{1+b_2y}$ for $(n_1,n_2)/(n_1,n_2)$	None	$\bar{x} = \frac{a_1x}{1+b_1x+b_2y}$ $\bar{y} = \frac{a_2y}{1+b_1x+b_2y}$ for n=m	$\bar{x} = \frac{a_1x}{1+b_1x+b_2y}$ $\bar{y} = \frac{a_2y}{1+b_1x+b_2y}$ for n=m

$$E_\star = \{(i,j) \in E \mid \text{the homogeneous system } d_{ij}=o \text{ has for a chosen normalization a unique solution}\} \text{ is symmetric} \qquad (II.7.2)$$

We call an approximant symmetric if for $F(x,y) = F(y,x)$ also $\frac{P_\star}{Q_\star}(x,y) = \frac{P_\star}{Q_\star}(y,x)$.

In the last row but one, one can find the variable changes $\bar{x}$ and $\bar{y}$ for which, if $\frac{P_\star}{Q_\star}(x,y)$ is the desired approximant for $F(x,y)$ then $\frac{P_\star}{Q_\star}(\bar{x},\bar{y})$ is the one for $F(\bar{x},\bar{y})$.

The HJA, LA and LAB^1 are denoted by $(n_1,n_2)/(m_1,m_2)$; for the CA $n_1=n_2=m_1=m_2=n$. The KWA are denoted by n/m. We remark also that the APA can be calculated recursively by means of the ε-algorithm and that they satisfy the consistency-property formulated in § 3. of this chapter, two properties which are for instance not satisfied by the Canterbury approximants.

7.3. *Numerical examples*

Let N_c be the number of unknown coefficients in the numerator and denominator of the approximant. For rational approximants 1 coefficient can always be determined by a normalization. We consider N_c-1 to be a measure for the operator-fitting ability of the calculated rational approximant.

For CA, HJA and LA: $N_c = (n_1+1)(n_2+1)+(m_1+1)(m_2+1)$.

For KWA and normalized APA: $N_c = [(n+1)(n+2)+(m+1)(m+2)]/2$.

a) Let us consider

$$F : \mathbb{R}^2 \to \mathbb{R} : \binom{x}{y} \to \frac{xe^x - ye^y}{x - y} = \sum_{i,j=0}^{\infty} \frac{1}{(i+j)!} x^i y^j .$$

In the Taylor series expansion of F we have a term in every power $x^i y^j$.

For KWA we have used the diagonal enumeration of points in $\mathbb{N}^2$, i.e. (o,o),(1,o),(o,1),(2,o),(1,1),(o,2),...

We compare the function values in some points. We see that the APA is good as well for $x > y$ as for $x < y$ (on a not too large neighbourhood of the origin), while other approximations, except CA(1,1)/(1,1), are not. The reason is simple: (1,1)/(1,0) fits the behaviour of F if $x > y$ and (1,1)/(0,1) fits the behaviour of F if $y > x$. The success of the (1,1) APA and the CA(1,1)/(1,1) partially lies in their conservation of the symmetry of F.

		N_c					
x			0.05	0.25	0.25	0.65	0.65
y			0.25	0.05	0.45	0.45	0.85
F(x,y)	$\frac{xe^x-ye^y}{x-y}$		1.342	1.342	1.924	2.697	3.718
$LAB^1(1,1)/(1,0)$	$\frac{1+\frac{1}{2}x+y}{1-\frac{1}{2}x}$	6	1.308	1.343	1.800	2.630	3.222
$LAB^1(1,1)/(1,1)$	$\frac{1+\frac{1}{2}(x+y)-\frac{1}{4}xy}{1-\frac{1}{2}(x+y)+\frac{1}{4}xy}$	8	1.328	1.328	2.032	2.109	4.153
(1,1)APA	$\frac{x+y+\frac{1}{2}(x^2+3xy+y^2)}{x+y-\frac{1}{2}(x^2+xy+y^2)}$	10	1.344	1.344	1.958	2.887	4.455
CA(1,1)/(1,1)	$\frac{1+\frac{1}{2}(x+y)-\frac{1}{6}xy}{1-\frac{1}{2}(x+y)+\frac{1}{3}xy}$	8	1.344	1.344	1.936	2.742	3.819
HJA(1,1)/(0,1)	$\frac{1+x+\frac{1}{2}y}{1-\frac{1}{2}y}$	6	1.343	1.308	1.903	2.419	3.609
KWA 1/1	$\frac{1+\frac{1}{2}x+y}{1-\frac{1}{2}x}$	6	1.308	1.343	1.800	2.630	3.222

b) Now consider

$$F : \mathbb{R}^2 \to \mathbb{R} : \binom{x}{y} \to \sqrt{1+x+y} = 1 + \frac{x+y}{2} + \sum_{k=2}^{\infty} (-1)^{k-1} \frac{(2k-3)!!}{2^k\, k!} (x+y)^k$$

where $(2k-3)!! = (2k-3).(2k-5)\ldots 5.3.1.$

We calculate some approximants. For the LA we also give the interpolationset E because the approximant depends on the chosen E.

The border of the domain of F is nicely simulated by the poles of the (k,1) APA:

$$y = -x - \frac{2k+2}{2k-1} \text{ with } \lim_{k\to\infty} \frac{-2k-2}{2k-1} = -1$$

(1,1) APA
$$\frac{1 + 0.75\ (x+y)}{1 + 0.25\ (x+y)}$$

CA (1,1)/(1,1)
$$\frac{1 + 0.75\ (x+y) - 0.1875\ xy}{1 + 0.25\ (x+y) - 0.1875\ xy}$$

HJA (1,1)/(1,0)
$$\frac{1 + 0.75x + 0.5y - 0.125\ xy}{1 + 0.25\ x}$$

HJA (1,1)/(0,1)
$$\frac{1 + 0.5x + 0.75y - 0.125\ xy}{1 + 0.25\ y}$$

KWA 1/1
$$\frac{1 + 0.75\ (x+y)}{1 + 0.25\ (x+y)}$$

LA (1,1)/(1,1)
$$\frac{1 + 0.75\ (x+y) - 0.1875\ xy}{1 + 0.25\ (x+y) - 0.1875\ xy}$$

 or 

LA (1,1)/(1,0)
$$\frac{1 + 0.75x + 0.5y - 0.125\ xy}{1 + 0.25\ x}$$

LA (1,1)/(0,1)
$$\frac{1 + 0.5x + 0.75y - 0.125\ xy}{1 + 0.25\ y}$$

We also compare the function values in some points and see that the (1,1) APA and the KWA 1/1 are much more accurate than the other types of approximants that have the same operator-fitting ability.

	(x,y)=(2,-1)	(x,y)=(-0.4,-0.5)	(x,y)=(2,-2)
F	1.4142	0.3162	1.0000
(1,1) APA, KWA 1/1	1.4000	0.4194	1.0000
CA(1,1)/(1,1), LA(1,1)/(1,1)	1.3077	0.3898	1.0000
HJA(1,1)/(1,0), LA(1,1)/(1,0)	1.5000	0.4722	1.3333
HJA(1,1)/(0,1), LA(1,1)/(0,1)	2.0000	0.4571	2.0000

c) Let us take a look at

$$F: \mathbb{R}^2 \to \mathbb{R}: \binom{x}{y} \to 1 + \frac{x}{0.1-y} + \sin(xy)$$

We calculate some approximants $\frac{P_\star}{Q_\star}(x,y)$, their operator-fitting ability and the exact order of $(F.Q_\star - P_\star)$.

If a Canterbury approximant is not uniquely determined we call it degenerate [30].
The interpolationset E prescribed for the calculation of $LAB^1(1,0)/(1,1)$, $LAB^1(1,2)/(0,2)$, $LAB^1(2,1)/(0,2)$, $LAB^1(2,2)/(1,1)$ always supplied a system of linearly dependent equations. So we do not include these approximants here.
Next to the type of the approximant one can find some small remarks. If several types provide the same rational function, they are grouped and then the multivariate Padé approximant is given after the small remarks.

We have also calculated an estimate ε_r of $\frac{\sup_A |F(x,y) - \frac{P_\star}{Q_\star}(x,y)|}{\sup_A |F(x,y)|}$ which is a measure for the relative error made by approximating ($\sup_A |F(x,y)| \simeq 10$). When we compare ε_r for the approximants $\frac{P_\star}{Q_\star}(x,y)$ that have the same operator-fitting ability, we remark that we can arrange them as follows from better to worse.

N_c		ε_r
6	HJA(1,1)/(0,1) and LA(1,1)/(0,1)	0.06
	(1,1)APA and KWA 1/1	0.09
	HJA(1,1)/(1,0)	0.73
	HJA(1,0)/(1,1) and LA(1,0)/(1,1)	0.81
	HJA(1,0)/(1,1)	90.1
8-9	HJA(1,2)/(0,2) and LA(1,2)/(0,2)	$0.9 \star 10^{-7}$
	(2,1) APA, KWA 2/1, CA(1,1)/(1,1), HJA(2,1)/(0,2) and LA(2,1)/(0,2)	0.06
	KWA 1/2	0.09
13-14	(3,1)APA, KWA 3/1, CA(2,2)/(1,1) and LA(2,2)/(1,1)	$0.9 \star 10^{-7}$
	(1,2) APA	0.07

Type	$\frac{P_\star}{Q_\star}(x,y)$		N_c	$F.Q_\star - P_\star$
HJA(1,1)/(1,0)	$1 + 10x + 101\ xy$		6	$O(xy^2)$
HJA(1,1)/(0,1)	degenerate		6	$O(xy^2)$
HJA(2,1)/(0,2)	degenerate		9	$O(xy^2)$
LA(1,1)/(0,1)	no interpolationset E supplying a unique approximant		6	$O(xy^2)$
	$\frac{1 + 10x + \alpha y + (101 + 10\alpha)\ xy}{1 + \alpha y}$	$\alpha = -\frac{1000}{101}$	—	$O(xy^3)$
CA (1,1)/(1,1)	degenerate		8	$O(x^2y, xy^2)$
	$\frac{1 + 10x + 10(1-\alpha)\ y + \alpha xy}{1 + 10(1-\alpha)\ y + (101\alpha - 201)\ xy}$	$\alpha = \frac{201}{101}$	—	$O(xy^3)$
LA(2,1)/(0,2) 	$\frac{1 + 10x - \frac{1000}{101}y + \frac{201}{101}xy}{1 - \frac{1000}{101}y}$		9	$O(xy^3)$
(2,1) APA			9	$O(xy^3)$
KWA 2/1			9	$O(xy^3)$

HJA(0,1)/(1,1)	degenerate $\dfrac{1 - (\frac{101+\alpha}{10})y}{1 - 10x - (\frac{101+\alpha}{10})y + \alpha xy}$	6	$O(x^2)$ for every α
HJA(1,0)/(1,1)	$\dfrac{1 + 10x}{1 - 101xy}$	6	$O(x^2y)$
LA(1,0)/(1,1) 		6	$O(x^2y)$
KWA 1/1	$\dfrac{1 + 10x - 10.1y}{1 - 10.1\,y}$	6	$O(xy^2)$
(1,1) APA		6	$O(xy^2)$
KWA 1/2		9	$O(xy^2)$
LA(1,2)/(0,2) 	no interpolationset E supplying a unique approximant	9	$O(xy^3)$
HJA(1,2)/(0,2)	degenerate $\dfrac{1+10x+\alpha y+(101+10\alpha)xy+\beta y^2+(10\beta+101\alpha+1000)xy^2}{1+\alpha y+\beta y^2}$	9	$O(xy^3)$
	$\alpha = -10,\ \beta = 0$	—	$O(x^3y^3)$

CA(2,2)/(1,1)	degenerate $P_\star(x,y) =$ $1+(10+\alpha)x-10y+(\beta+1)xy+10\alpha x^2-10xy^2+$ $(10\beta+101\alpha)x^2y+(101\beta+1000\alpha)x^2y^2$ $Q_\star(x,y) = 1+\alpha x-10y+\beta xy$	13	$O(x^2y^3)$
	$\alpha = 0,\ \beta = 0$	—	$O(x^3y^3)$
KWA 3/1 (3,1)APA	$\frac{1+10x-10y+xy-10xy^2}{1-10y}$	13 13	$O(x^3y^3)$ $O(x^3y^3)$
(1,2)APA	$\frac{x-1.01y+10y^2+10x^2-20.2xy}{x-1.01y+10y^2-10.1xy+2.01xy^2}$	14	$O(xy^4,x^2y^3)$
LA(2,2)/(1,1)	no interpolationset E supplying a unique approximant $P_\star(x,y) =$ $1+(10+\alpha)x+10\alpha x^2-10y+(1-10\alpha)xy$ $+\alpha x^2y-10xy^2-10\alpha x^2y^2$ $Q_\star(x,y) = 1+\alpha x-10y-10\alpha xy$	13	$O(x^3y^3)$ for every α

Remark the fact that the accuracy of Canterbury and Lutterodt approximants depends mainly on the chosen type of approximant, i.e. on the degrees of x and y in numerator and denominator; one can obtain a very accurate or a very bad approximant with the same amount of work, because one cannot always tell from the first Taylor coefficients of F which degrees one should choose. And most of the times the only information one gets about the multivariate function are some Taylor coefficients. If the denominator of the Padé approximant equals 1-10y, then the rational function has the same poles as the given multivariate function F and the only remaining terms in $(F.Q_\star - P_\star)$ come from sin(xy). This explains the fact that ε_r diminishes tremendously for certain types of approximants $(0.9 \star 10^{-7})$.

7.4. *Rational approximations of multiple power series introduced by Hillion* [28]

He also only considers double series because the extension to many variables is straigh forward. We briefly repeat his definition of rational approximations.

Given the double series

$$F(x,y) = \sum_{i,j=0}^{\infty} c_{ij} x^i y^j$$

we introduce the polynomials

$$\emptyset_{k,p}(x,y) = \begin{cases} \sum_{\ell=0}^{k-p} c_{k-\ell,p+\ell}\, x^{k-\ell} y^{p+\ell} + \sum_{\ell=0}^{p-1} (c_{k,\ell}\, x^k y^\ell + c_{\ell,k}\, x^\ell y^k) & \text{if } k > p \\ c_{k,k}\, x^k y^k + \sum_{\ell=0}^{k-1} (c_{k,\ell}\, x^k y^\ell + c_{\ell,k}\, x^\ell y^k) & \text{if } k \le p \end{cases}$$

It is easy to see that for p fixed

$$\sum_{k=0}^{\infty} \emptyset_{k,p}(x,y) = F(x,y)$$

The rational approximation $[n/m]_p(x,y)$ is now defined by the ε-algorithm

$$\varepsilon_{-1}^{(i,p)} = 0 \qquad i = 0,1,\ldots$$

$$\varepsilon_{0}^{(i,p)} = \sum_{k=0}^{i} \emptyset_{k,p}(x,y) \qquad i = 0,1,\ldots$$

$$\varepsilon_{j+1}^{(i,p)} = \varepsilon_{j-1}^{(i+1,p)} + \frac{1}{\varepsilon_j^{(i+1,p)} - \varepsilon_j^{(i,p)}} \qquad \begin{matrix} j = 0,1,\ldots \\ i = 0,1,\ldots \end{matrix}$$

$$[n/m]_p(x,y) = \varepsilon_{2m}^{(n-m,p)}$$

If we take p=o we obtain precisely the (n,m) APA. The applicability of the -algorithm for the calculation of the (n,m) APA was proved in § 7. of chapter I.

§ 8. BETA FUNCTION

8.1. *Introduction*

The Beta function is an example which has also been studied by the Canterbury group [25] and by Levin [35]. We will compare our results with theirs. The Beta function may be defined by

$$B(x,y) = \frac{\Gamma(x)\,\Gamma(y)}{\Gamma(x+y)}$$

where Γ is the Gamma function. Singularities occur for $x = -k$ and $y = -k$ ($k=0, 1, 2, \ldots$ and zeros for $y = -x-k$ ($k=0, 1, 2, \ldots$).

We write

$$B(x,y) = \frac{A(x-1,y-1)}{xy}$$

with

$$A(u,v) = 1 + uv\, f(u,v)$$

The coefficients in the Taylor series expansion of $f(u,v)$ have been calculated by the first method suggested in [25]. We will calculate some (n,m) APA $\frac{P}{Q}(u,v)$ for $f(u,v)$ and compute

$$\frac{1 + (x-1)(y-1)\frac{P}{Q}(x-1,y-1)}{xy}$$

as an approximation for $B(x,y)$. Also we will compare the singularities and zeros of $[1 + (x-1)(y-1)\frac{P}{Q}(x-1,y-1)]/xy$ with those of $B(x,y)$. The numerical values of the APA can easily be calculated via the ε-algorithm, while the coefficients in numerator and denominator can be calculated by solving a linear system whose matrix has low displacement-rank.

Let us first take a look at the computational effort it takes for the calculation of a certain approximant. We denote by N_f the number of coefficients in the Taylor series of f which we shall need for the computation of the approximant; N_u still denotes the number of unknown coefficients in the homogeneous system.

For a HJA(n,n)/(m,m):

$$N_u = (m+1)^2$$
$$N_f = (m+1)^2 + (n+1)^2 + 2\min(n,m) - 1$$

For an (n,m) APA:

for $nm > 0$: $N_u = [(nm+m+1)(nm+m+2) - nm(nm+1)]/2$

for $nm = 0$: $N_u = (m+2)(m+1)/2$

$$N_f = (n+m+1)(n+m+2)/2$$

The rational functions which Levin used for the approximation of the Beta function, were of the following type

$$\frac{\displaystyle\sum_{j=0}^{n_1} x^j \frac{\sum_{i=0}^{n_2} \alpha_{ij} y^i}{\sum_{i=0}^{n_2} \beta_{ij} y^i} + \sum_{j=0}^{n_1} y^j \frac{\sum_{i=0}^{n_2} p_{ij} x^i}{\sum_{i=0}^{n_2} q_{ij} x^i}}{\displaystyle\sum_{i=0}^{m} \sum_{j=0}^{m} a_{ij} x^i y^j}$$

and we shall denote them by $[(n_1;n_2)/m]_r$ because for their computation:

$$N_u = (m+1)^2 + (n_2+1)(n_1+1)$$
$$N_f = 2(2n_2+1)(n_1+1) - (n_1+1)^2 + [\max(0,m+r-n_1)]^2 - 1$$

(for more details see [35]).

Using the prong-method [30] the homogeneous system of equations for the calculation of HJA(n,n)/(m,m) can be solved in $O[m^2(2m^2+2m-1)]$ operations.

Exploiting the fact that the matrix of the homogeneous system of equations has low displacement-rank $\alpha(H)$, the denominator of the (n,m) APA can be calculated in $O(\alpha(H)N_e^2)$, so at most in $O[\frac{m+2}{4}((nm+n+m+2)(nm+n+m+1)-(nm+n+2)(nm+n+1))^2]$ operations.

But the calculation of a function value of the (n,m) APA can via the ε-algorithm already be performed in $O[(n+m)^2+m^2]$ operations and we prefer this method to the solution of the system.

The solution of the homogeneous system for the calculation of $[(n_1;n_2)/m]_r$ involves $O[(m+1)^6 + (n_2+1)^2(n_1+1)]$ operations because each system in the q_{ij} has a Toeplitz structure.

After comparison of the N_f, N_u and the computational effort we decided to compare (see also [35]) the numerical values of

(8,4)APA with $[(4;5)/2]_3$ and HJA(7,7)/(3,3)

(4,4)APA with $[(3;3)/1]_1$ and CA(3,3)/(3,3)

(8,3)APA with $[(2;5)/2]_2$ and HJA(7,7)/(2,2)

We shall also give the trajectories of the poles and zeros of some Canterbury approximants and some abstract Padé approximants (Levin did not draw any figures illustrating the situation of poles and zeros).

It is easy to see that the APA can produce better results than the HJA and the CA, e.g. for (x,y)=(-0.75,-0.75), and that they can also produce better results than the approximants Levin used, e.g. for (x,y)=(0.50,0.50). They are most accurate for (u,v)=(x-1,y-1) not too far from the origin.

8.2. *Numerical values*

We compare the numerical values of the approximant with the exact values of B(x,y) at various points.

Table II.8.1.

(x,y)	(-0.75,-0.75)	(-0.50,-0.50)	(-0.25,-0.25)	(0.25,0.25)	(0.50,0.50)	(0.75,0.75)	(-1.75,1.75)
B(x,y)	9.88839829	0.	-6.77770467	7.41629871	3.14159265	1.694426166	0.
$[(4;5)/2]_3$	9.888	-0.00021	-6.777755	7.41629594	3.14159248	1.69442616	0.0186
HJA(7,7)/(3,3)	9.820	-0.0010	-6.77774	7.41629871	3.14159265	1.69442617	0.0016
(8,4)APA	9.884	-0.00006	-6.777705	7.41629871	3.14159265	1.69442617	-0.0351
(x,y)	(-0.75,-0.75)	(-0.50,-0.50)	(-0.25,-0.25)	(0.25,0.25)	(0.50,0.50)	(0.75,0.75)	(0.75,0.25)
B(x,y)	9.88839829	0.	-6.77770467	7.41629871	3.14159265	1.694426166	4.44288293
$[(3;3)/1]_1$	9.94	-0.03	-6.794	7.416229	3.14159242	1.69442617	4.442883
CA(3,3)/(3,3)	7.0	-0.14	-6.787	7.416310	3.14159269	1.69442617	4.442883
(4,4)APA	8.38	-0.13	-6.802	7.416281	3.14159263	1.69442617	4.442883
(x,y)	(-0.75,-0.75)	(-0.50,-0.50)	(-0.25,-0.25)	(0.25,0.25)	(0.50,0.50)	(0.75,0.75)	(1.75,-0.75)
B(x,y)	9.88839829	0.	-6.77770467	7.41629871	3.14159265	1.694426166	-4.44288293
$[(2;5)/2]_2$	9.86	-0.003	-6.7783	7.41629639	3.14159252	1.69442617	-4.4428
HJA(7,7)/(2,2)	9.3	-0.014	-6.7783	7.41629881	3.14159265	1.69442617	-4.4421
(8,3)APA	9.74	-0.006	-6.7783	7.41629862	3.14159265	1.69442617	-4.4442

8.3. *Figures*

The pattern of singularities and zeros of the Beta function B(x,y) itself is shown in figure II.8.1.

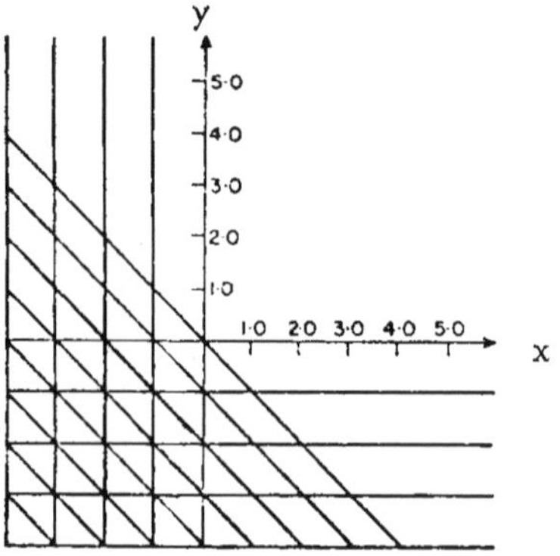

Figure II.8.1.

The situation of poles and zeros of CA(2,2)/(2,2) and HJA(7,7)/(2,2) is illustrated in the figures II.8.2 and II.8.3 respectively. The poles and zeros of (0,2)APA, (2,2)APA and (7,1)APA are drawn in the figures II.8.4, II.8.5a-b and II.8.6a-b respectively. In both cases we remark that the vertical, horizontal and diagonal lines are nicely simulated.

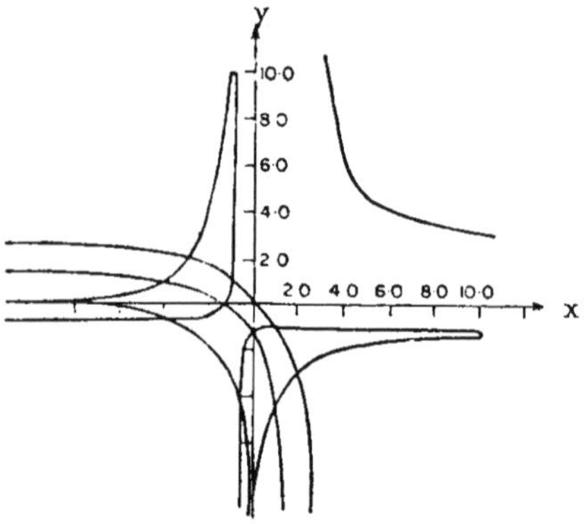

Figure II.8.2.

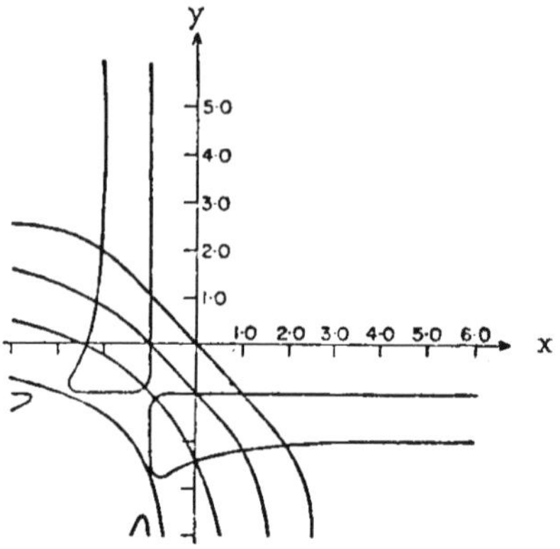

Figure II.8.3.

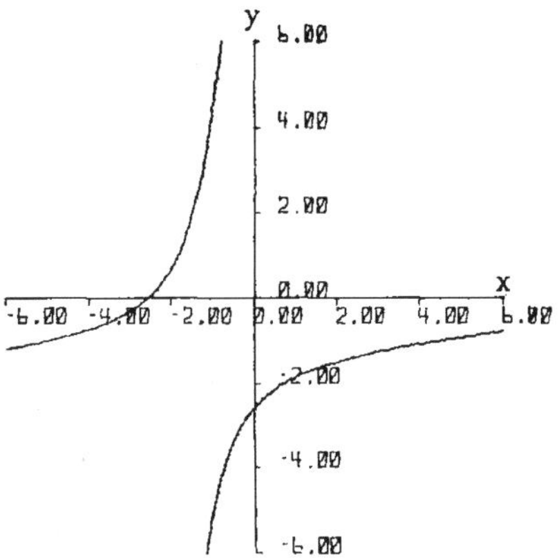

Figure II.8.4.: poles

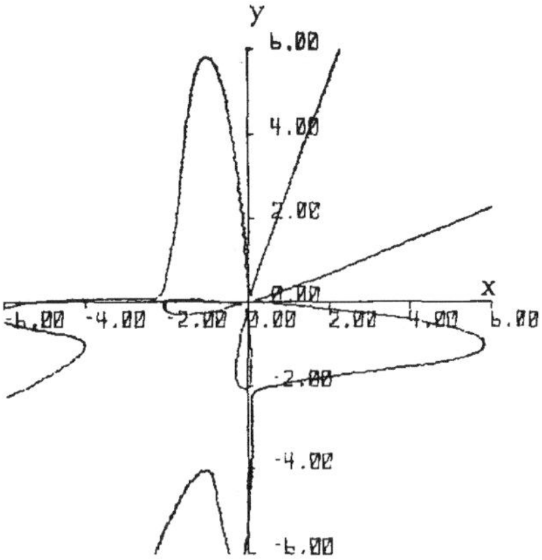

Figure II.8.5a.: poles

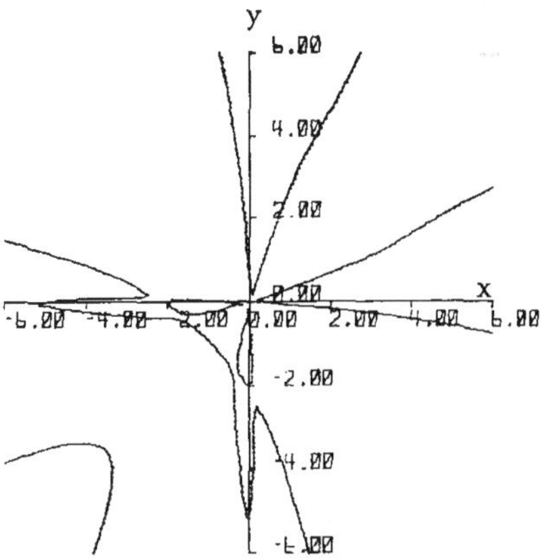

Figure II.8.5b.: zeros

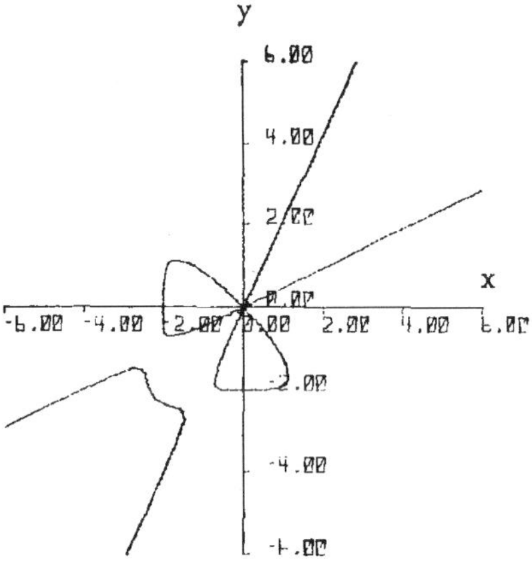

Figure II.8.6a.: poles

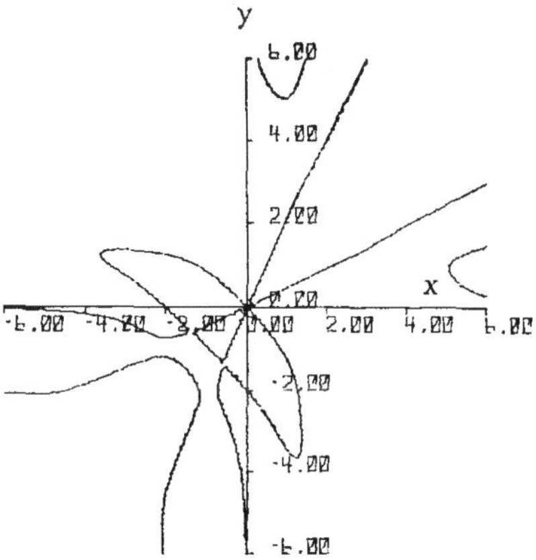

Figure II.8.6b.: zeros

CHAPTER III: THE SOLUTION OF NONLINEAR OPERATOR EQUATIONS

§ 1. INTRODUCTION

Several types of nonlinear operator equations

$$F(x) = 0$$

will be considered. Iterative methods for the solution of those operator equations are introduced and discussed in § 2. and § 3. Starting from an approximation x_0 for a root $x^\star$ of F, a sequence of further approximations $\{x_i\}$ is constructed in such a way that x_{i+1} is computed by means of x_i. The well-known Newton- and Chebyshev-iteration [41 pp. 205] are special cases. Among others, an interesting new iterative procedure which we shall call the Halley-iteration, is constructed.
Afterwards systems of nonlinear equations, initial value problems, boundary value problems, partial differential equations and nonlinear integral equations are respectively treated in the paragraphs 4, 5, 6, 7 and 8. We will remark that in the neighbourhood of singularities iterative procedures that are derived from solutions of the Padé approximation problem of order (n,m) with m > o (cfr. Halley's method) are more suitable than those where m = o. Finally the numerical stability of the Halley-iteration for the solution of a system of nonlinear equations will be discussed in paragraph 9.

§ 2. INVERSE INTERPOLATION

Consider the nonlinear operator $F: X \to Y$ where again X is a Banach space and Y is a commutative Banach algebra. Suppose we want to find $x^\star$ in X such that

$$F(x^\star) = 0$$

Let F be abstract analytic in a neighbourhood U of $x^\star$ and let $x^\star$ be a simple root of F, in other words let $F'(x^\star)^{-1}$ exist and be a bounded linear operator. Then there is a neighbourhood V of 0 such that the inverse operator $G: V \subset Y \to U \subset X$ exists and is abstract analytic in V [6 pp. 299-301].
By means of solutions of the Padé approximation problem for the inverse operator G (X must be a commutative Banach algebra then), we can construct iterative methods to find $x^\star$ (inverse interpolation).
By F_i' and F_i'' we mean respectively the first and second Fréchet-derivative of F at x_i. Let $F_i = F(x_i) = y_i$ and $G(y_i) = x_i$. We know that $G(0) = x^\star$ and that G is analytic in a neighbourhood of 0; so we can write [41 pp. 205]

$$G(y) = G(y_i) + F_i'^{-1}(y-y_i) - \frac{1}{2}F_i'^{-1}(F_i'' F_i'^{-1})(y-y_i)^2 + \dots \qquad \text{(III.2.1)}$$

where $(F_i'' F_i'^{-1})(y-y_i)^2$ is the bilinear operator F_i'' evaluated in $(F_i'^{-1}(y-y_i), F_i'^{-1}(y-y_i))$.
If we calculate a solution (P_i, Q_i) of the Padé approximation problem of order (n,m) for G in y_i, we could iterate

$$x_{i+1} = (\frac{1}{Q_i}.P_i)(0) \text{ or } (\frac{1}{Q_{*i}}.P_{*i})(0)$$

where $\frac{1}{Q_{*i}}.P_{*i}$ is a reduced rational form of $\frac{1}{Q_i}.P_i$.

Observe that the well-known Newton-iteration results from approximating the series (III.2.1) by its first two terms, i.e. a solution of the Padé approximation problem of order (1,0) for G:

$$x_{i+1} = x_i + a_i \text{ where } a_i = -F_i'^{-1}F_i \tag{III.2.2}$$

The (0,1) Padé approximation problem gives the following iterative method:

$$x_{i+1} = x_i^2/(x_i - a_i) \tag{III.2.3}$$

where the multiplication and division are those in the commutative Banach algebra X. The first three terms in (III.2.1), which form in fact a solution of the (2,0) Padé approximation problem, could also be used to approximate x^*, giving the next iteration:

$$x_{i+1} = x_i + a_i - \frac{1}{2}F_i'^{-1}F_i''a_i^2 \tag{III.2.4}$$

The iteration (III.2.4) is known as Chebyshev's method for the solution of operator equations.

Another way to approximate x^* is to use a solution of the (1,1) Padé approximation problem for the series in (III.2.1):

$$x_{i+1} = x_i + \frac{a_i^2}{a_i + \frac{1}{2}F_i'^{-1}F_i''a_i^2} \tag{III.2.5}$$

which is a generalization of a formula of Frame [18] and a rediscovery of the Halley-correction, now for operator equations.

If $F_i'^{-1}F_i''a_i^2 = a_i.La_i$ for a bounded linear operator L, then (III.2.5) reduces to:

$$x_{i+1} = x_i + \frac{a_i}{I + \frac{1}{2}La_i}$$

where I is now the unit for the multiplication in the Banach algebra X. If $X = \mathbb{R} = Y$ this reduction can always be performed and (III.2.5) then results in the classical Halley-iteration. The iterative procedure (III.2.5) is closely related to the method of tangent hyperbolas [39 pp. 188]:

$$x_{i+1} = x_i - \{F_i' + \frac{1}{2}F_i''a_i\}^{-1} F_i$$

which can also be written as

$$x_{i+1} = x_i + \{I_X + \frac{1}{2}F_i'^{-1}F_i''a_i\}^{-1} a_i$$

where I_X: X → X: x → x is the identity. This second formulation shows the interrelation with (III.2.5): the operator $\{I_X + \frac{1}{2}F_i'^{-1}F_i''a_i\}$ is evaluated in a_i, the vector a_i is multiplied by a_i and those two vectors are divided in order to avoid the inversion of $\{I_X + \frac{1}{2}F_i'^{-1}F_i''a_i\}$. This technique is similar to a method introduced by Altman to avoid the inversion of matrices in a procedure to solve a system of nonlinear equations.

One of the main drawbacks to the use of (n,m) Padé approximants is the computational cost of evaluating higher derivatives of F. However, in some cases these derivatives can be computed quite easily, e.g. if F satisfies a certain differential equation (so

that the derivatives can be computed from this equation rather than from F itself) or if F is a composition of polynomial, trigonometric or exponential functions.
Let us now suppose that the iterative procedure chosen for the calculation of a simple root $x^\star$ of F is convergent, i.e. $\lim_{i\to\infty} x_i = x^\star$ or equivalently $\lim_{i\to\infty} \|x_i - x^\star\| = o$.

Definition III.2.1.:

An iterative procedure which calculates x_{i+1} by means of x_i is of <u>order p</u> if for all i, there exist integers $p_1 \geq o$ and $p_2 > o$ and there exist multilinear operators

$$E_{p_1} \in L(X^{p_1}, Y) \text{ and } E_{p_2} \in L(X^{p_2}, Y) \text{ with}$$

$$E_{p_1} (x-x_i)^{p_1} \not\equiv 0 \text{ such that}$$

$$[E_{p_1}(x^\star-x_i)^{p_1}].(x^\star-x_{i+1}) = E_{p_2}(x^\star-x_i)^{p_2} \text{ and } p = p_2-p_1.$$

In classical definitions of order of an iterative process, the factor $E_{p_1}(x^\star-x_i)^{p_1}$ on the left hand side is missing.
Its presence here is due to $\partial_o P_i$ and $\partial_o Q_i$ or $\partial_o P_{\star i}$ and $\partial_o Q_{\star i}$ in the abstract Padé approximation problem; this will be made clearer in the next theorem. Nevertheless this definition is an extension of the well-known definition [38 pp. 148] because for p_1=o and E_o regular in X we can prove that there exist J_i in $\mathbb{R}_o^+$ such that

$$\|x^\star - x_{i+1}\| \leq J_i \|x^\star - x_i\|^p$$

We will now use the notation $\frac{1}{Q_{\star i}}.P_{\star i}$ for a representant of the rational operators that can be formed with the elements (P_i,Q_i) and $(P_{\star i},Q_{\star i})$ of the (n,m) abstract Padé approximant for G in y_i, which is an equivalence class.

Theorem III.2.1.:

The order of the iterative procedure $x_{i+1} = (\frac{1}{Q_{\star i}}.P_{\star i})(0)$ is at least $nm+n+m+1-\partial_o Q_i$ if $D(T_{t_o}) \neq \emptyset$ where T is such that $P_i = P_{\star i}.T$, $Q_i = Q_{\star i}.T$ and $t_o = \partial_o T$.

Proof:

Because of theorem I.5.4 we can write

$$(G.Q_{\star i} - P_{\star i})(y) = O((y-y_i)^{nm+n+m+1-t_o})$$

$$\text{where } Q_{\star i}(y) = \sum_{j=\partial_o Q_{\star i}}^{\partial Q_{\star i}} B_{\star j}(y-y_i)^j$$

For $y = 0$ we have, since $G(0) = x^\star$ and $x_{i+1} = (\frac{1}{Q_{\star i}}.P_{\star i})(0)$:

$$Q_{\star i}(0) \cdot (x^\star - x_{i+1}) = (G.Q_{\star i} - P_{\star i})(0).$$

Let $p_2 = \partial_o(G.Q_{\star i} - P_{\star i}) \geq nm+n+m+1-t_o$ and $p_1 = \partial_o Q_{\star i}$.

Since $x^\star$ is a simple root, G is sufficiently differentiable in a neighbourhood of 0 containing the line segment joining the points 0 and y_i and so via Taylor's theorem [41 pp. 124]

$$(G.Q_{\star i} - P_{\star i})(0) = \int_0^1 \frac{(1-\theta)^{p_2-1}}{(p_2-1)!} (G.Q_{\star i} - P_{\star i})^{(p_2)} ((1-\theta)y_i)(-y_i)^{p_2}\, dt$$

$$= D_{p_2} (-y_i)^{p_2}$$

$$Q_{\star i}(0) = \int_0^1 \frac{(1-\theta)^{p_1-1}}{(p_1-1)!} Q_{\star i}^{(p_1)} ((1-\theta)y_i)(-y_i)^{p_1}\, dt$$

$$= D_{p_1} (-y_i)^{p_1}$$

for certain multilinear operators

$D_{p_1} \in L(Y^{p_1}, X)$ and $D_{p_2} \in L(Y^{p_2}, X)$.

$$\text{Now } -y_i = F(x^\star) - F(x_i)$$
$$= \{\int_0^1 F'(\theta x^\star + (1-\theta)\, x_i)\, d\theta\}\, (x^\star - x_i)$$
$$= L(x^\star - x_i)$$

with L a linear operator and thus

$$[E_{p_1} (x^\star - x_i)^{p_1}].(x^\star - x_{i+1}) = E_{p_2} (x^\star - x_i)^{p_2}$$

with $E_{p_1} (x^\star - x_i)^{p_1} = D_{p_1} (L(x^\star - x_i))^{p_1}$ and

$E_{p_2} (x^\star - x_i)^{p_2} = D_{p_2} (L(x^\star - x_i))^{p_2}$.

If we write $p = p_2 - p_1$ then $p \geq nm+n+m+1-t_o-\partial_o Q_{\star i} = nm+n+m+1-\partial_o Q_i$ because $\partial_o Q_{\star i} + t_o = \partial_o Q_i$.

■

Using theorem III.2.1 we see that

Newton's method has order 2

iteration (III.2.3) has order 2

Chebyshev's method has order 3

Halley's method has order 3

According to definition III.2.1 the method of tangent hyperbolas is also of order 3.

§ 3. DIRECT INTERPOLATION

Since F is analytic in a neighbourhood of $x^{\star}$ containing the approximants x_i, we can approximate F by $\frac{1}{Q_i}\cdot P_i$ or $\frac{1}{Q_{\star i}}\cdot P_{\star i}$ where (P_i,Q_i) is a solution of the Padé approximation problem of order (n,m) for F in x_i. We then calculate x_{i+1} such that $P_i(x_{i+1}) = 0$ or $P_{\star i}(x_{i+1}) = 0$ and iterate (direct interpolation).

Let us again take a look at the iterative procedures we obtain if $n+m \leq 2$ and $n > 0$.

First of all we write down the Taylor series expansion

$$F(x) = F(x_i) + F_i'(x-x_i) + \frac{1}{2}F_i''(x-x_i)^2 + \dots \qquad \text{(III.3.1)}$$

The use of the (1,0) Padé approximation problem gives

$$F_i + F_i'(x_{i+1} - x_i) = 0$$

or equivalently

$$x_{i+1} = x_i - F_i'^{-1}F_i$$

which is precisely Newton's method.

When we use a solution of the (2,0) Padé approximation problem we obtain

$$F_i + F_i'(x_{i+1}-x_i) + \frac{1}{2}F_i''(x_{i+1}-x_i)^2 = 0$$

so that we have to solve a quadratic operator equation. As indicated in [42], solving such an equation is a quite complicated matter; moreover, the choice of x_{i+1} among distinct solutions of the quadratic equation is also a problem.

However, an approximate solution $\overline{x}_{i+1}$ can be obtained in the following way [16].

The root of the quadratic equation satisfies

$$x_{i+1} = x_i - F_i'^{-1}F_i - \frac{1}{2}F_i'^{-1}F_i''(x_{i+1}-x_i)^2$$

If in the righthand side $x_{i+1}-x_i$ is approximated by the Newton-correction a_i, we have an approximation for x_{i+1} which is precisely a Chebyshev-iterationstep

$$\overline{x}_{i+1} = x_i + a_i - \frac{1}{2}F_i'^{-1}F_i''a_i^2$$

Another way to express x_{i+1} is

$$x_{i+1} = x_i - \{F_i' + \frac{1}{2}F_i''(x_{i+1}-x_i)\}^{-1}F_i$$

If again in the righthand side $x_{i+1}-x_i$ is approximated by a_i [15] we get

$$\overline{x}_{i+1} = x_i - \{F_i' + \frac{1}{2}F_i''a_i\}^{-1}F_i$$

which is the method of tangent hyperbolas.

A solution of the (1,1) Padé approximation problem for F in x_i is

$$(P_i,Q_i) = (F_iF_i'(x-x_i)+[F_i'(x-x_i)]^2 - \frac{1}{2}F_iF_i''(x-x_i)^2\ ,\ F_i'(x-x_i)-\frac{1}{2}F_i''(x-x_i)^2)$$

where the multiplication is now the one defined in the Banach algebra Y.
If x_{i+1} is such that $P_i(x_{i+1}) = 0$, we have to solve

$$F_i + F_i'(x_{i+1}-x_i) = \frac{1}{2}\frac{F_i \cdot F_i''(x_{i+1}-x_i)^2}{F_i'(x_{i+1}-x_i)}$$

If we approximate in the righthand side $x_{i+1}-x_i$ by a_i we get the approximate solution

$$\bar{x}_{i+1} = x_i + a_i - \frac{1}{2}F_i'^{-1}F_i''a_i^2$$

which is again Chebyshev's method.
If $F_i''(x-x_i)^2 = (F_i' \otimes L)(x-x_i)^2$ for a certain linear operator L, then $\frac{1}{Q_i}.P_i$ can be reduced to

$$(\frac{1}{Q_{\star i}}.P_{\star i})(x) = \frac{F_i + F_i'(x-x_i) - \frac{1}{2}F_i \otimes L(x-x_i)}{I - \frac{1}{2}L(x-x_i)}$$

Remark again that for $X = \mathbb{R} = Y$ this reduction can always be performed. If x_{i+1} is such that $P_{\star i}(x_{i+1}) = 0$, then

$$x_{i+1} = x_i - \{F'_i - \frac{1}{2}F_i \otimes L\}^{-1} F_i$$

where now $(F_i' - \frac{1}{2}F_i \otimes L) = F_i' + \frac{1}{2}(F_i'a_i) \otimes L = F_i' + \frac{1}{2}F_i''a_i$. So we have again the method of tangent hyperbolas.
We must conclude that the methods derived by direct interpolation are either too complicated (when we calculate the exact solution x_{i+1}) or similar to methods of § 2. (when we calculate an approximate solution $\bar{x}_{i+1}$). This justifies the fact that we will only use iterative procedures from § 2. for the solution of the different nonlinear operator equations.

§ 4. SYSTEMS OF NONLINEAR EQUATIONS

If we want to solve a system of p nonlinear equations in p real variables

$$F(x) = \begin{pmatrix} f_1(x_1, \ldots, x_p) \\ \vdots \\ f_p(x_1, \ldots, x_p) \end{pmatrix} = 0$$

then $X = \mathbb{R}^p = Y$ and the multiplication in X and Y is performed component-wise with $I=(1,\ldots,1)$ in $\mathbb{R}^p$. The successive approximations x_i in an iterative procedure are vectors in $\mathbb{R}^p$. The operator F_i' is represented by the Jacobian matrix

$$F_i' = \begin{pmatrix} \frac{\partial f_1(x)}{\partial x_1} & \frac{\partial f_1(x)}{\partial x_2} & \cdots & \frac{\partial f_1(x)}{\partial x_p} \\ \vdots & & & \\ \frac{\partial f_p(x)}{\partial x_1} & & \cdots & \frac{\partial f_p(x)}{\partial x_p} \end{pmatrix}\Bigg|_{x = x_i}$$

and the operator F_i'' by the hypermatrix

$$F_i'' = \left(\begin{array}{ccc|ccc|c|ccc} \frac{\partial^2 f_1(x)}{\partial x_1^2} & \cdots & \frac{\partial^2 f_1(x)}{\partial x_1 \partial x_p} & \frac{\partial^2 f_1(x)}{\partial x_2 \partial x_1} & \cdots & \frac{\partial^2 f_1(x)}{\partial x_2 \partial x_p} & \cdots & \frac{\partial^2 f_1(x)}{\partial x_p \partial x_1} & \cdots & \frac{\partial^2 f_1(x)}{\partial x_p^2} \\ \vdots & & \vdots & & & & & \vdots & & \vdots \\ \frac{\partial^2 f_p(x)}{\partial x_1^2} & \cdots & \frac{\partial^2 f_p(x)}{\partial x_1 \partial x_p} & & \cdots & & & \frac{\partial^2 f_p(x)}{\partial x_p \partial x_1} & & \frac{\partial^2 f_p(x)}{\partial x_p^2} \end{array} \right)\Bigg|_{x = x_i}$$

with $\frac{\partial^2 f_\ell(x)}{\partial x_j \partial x_k} = \frac{\partial^2 f_\ell(x)}{\partial x_k \partial x_j}$ for $j, k, \ell = 1, \ldots, p$.

Let us compare the numerical effort per iterationstep for the different iterative procedures.

Iteration (III.2.3) and Newton's method both solve one system of linear equations

$$F_i' \, a_i = -F_i$$

and combine x_i and a_i to find x_{i+1}.

Chebyshev's method, Halley's method and the method of tangent hyperbolas each solve two systems of linear equations

Chebyshev:
$$\begin{cases} F_i' \, a_i = -F_i \\ F_i' \, b_i = F_i'' \, a_i^2 \end{cases}$$
$$x_{i+1} = x_i + a_i - \frac{1}{2} b_i$$

Halley
$$\begin{cases} F_i' \, a_i = -F_i \\ F_i' \, b_i = F_i'' \, a_i^2 \end{cases}$$
$$x_{i+1} = x_i + \frac{a_i^2}{a_i + \frac{1}{2} b_i}$$

Tangent hyperbolas:
$$\begin{cases} F_i' \, a_i = -F_i \\ (F_i' + \frac{1}{2} F_i'' \, a_i) \, b_i = -F_i \end{cases}$$
$$x_{i+1} = x_i + b_i$$

However, for the first two methods these systems have the same coefficient matrix F_i' so that the elimination part of the Gauss-method has only to be performed once, while the third method requires the solution of linear systems with matrices F_i' and $F_i' + \frac{1}{2} F_i'' a_i$ so that the entire Gauss-method has to be performed twice. If we use the ε-algorithm for the calculation of the next iterationstep in Halley's method, we also have to solve two linear systems of equations:

$$\varepsilon_0^{(o)} = x_i \qquad \varepsilon_1^{(o)} = a_i^{-1}$$

$$\varepsilon_0^{(1)} = x_i + a_i \qquad \varepsilon_2^{(o)} = x_i + a_i - [a_i^{-1} + 2b_i^{-1}]^{-1} = x_{i+1}$$

$$\varepsilon_1^{(1)} = -2\, b_i^{-1}$$

$$\varepsilon_0^{(2)} = x_i + a_i - \frac{1}{2} b_i$$

Let us now compare the numerical results for the solution of a system of nonlinear equations where the inverse operator G has singularities in the neighbourhood of 0. Consider

$$F : \mathbb{R}^2 \to \mathbb{R}^2 : \binom{x}{y} \to \begin{pmatrix} \exp(-x+y)-0.1 \\ \exp(-x-y)-0.1 \end{pmatrix}$$

which has a simple root

$$x^\star = \begin{pmatrix} -\ell n\ (0.1) \\ 0 \end{pmatrix} = \begin{pmatrix} 2.302585092994046 \\ 0. \end{pmatrix}$$

The inverse operator

$$G : \mathbb{R}^2 \to \mathbb{R}^2 : \binom{u}{v} \to \begin{pmatrix} -\dfrac{\ell n(u+0.1)+\ell n(v+0.1)}{2} \\ \dfrac{\ell n(u+0.1)-\ell n(v+0.1)}{2} \end{pmatrix}$$

has singularities for $u=-0.1$ or $v=-0.1$.

In table III.4.1 one finds the consecutive iterationsteps of Newton's method and iteration (III.2.3) both of order 2 with $x_0 = (5.3, 0.3)$ as initial point. After 13 iterationsteps method (III.2.3) converges ($\|x_{13} - x_{12}\| \le 10^{-5}$) while Newton's method needs 28 iterationsteps to obtain the same accuracy.

In table II.4.2 one finds the results obtained by Halley's method and the method of tangent hyperbolas both of order 3 with $x_0 = (4.3, 2.0)$ as initial point. If Chebyshev's method is used, starting from the same initial point x_0, then the sequence of iterands diverges.

Clearly methods derived from rational approximations, like Halley's method and iteration (III.2.3), behave better in this case than methods derived from polynomial approximations, like Chebyshev's and Newton's method. The choice of the initial point also plays an important role: if it is close to the singularity, linear methods get into trouble, and if it is not, linear and rational methods can behave equally well.

	i	x_i	
Newton	0	0.53000000 (+01)	0.30000000 (+00)
	1	-0.14641978 (+02)	-0.58006624 (+01)
	2	-0.13641986 (+02)	-0.58006552 (+01)
	3	-0.12642005 (+02)	-0.58006355 (+01)
	4	-0.11642059 (+02)	-0.58005821 (+01)
	5	-0.10542204 (+02)	-0.58004369 (+01)
	6	-0.96425985 (+01)	-0.58000422 (+01)
	7	-0.86436705 (+01)	-0.57989703 (+01)
	8	-0.76465781 (+01)	-0.57960627 (+01)
	9	-0.56544360 (+01)	-0.57882050 (+01)
	10	-0.56754628 (+01)	-0.57671785 (+01)
	11	-0.47302660 (+01)	-0.57123764 (+01)
	12	-0.38637718 (+01)	-0.55788736 (+01)
	13	-0.31416378 (+01)	-0.53010155 (+01)
(III.2.3)	0	0.53000000 (+01)	0.30000000 (+00)
	1	0.11128288 (+01)	0.14061045 (-01)
	2	0.29686569 (+01)	0.10780485 (-01)
	3	0.22508673 (+01)	0.36586016 (-02)
	4	0.23024184 (+01)	0.18765919 (-02)
	5	0.23025834 (+01)	0.93837391 (-03)
	6	0.23025847 (+01)	0.46918733 (-03)
	7	0.23025850 (+01)	0.23459371 (-03)
	8	0.23025851 (+01)	0.11729686 (-03)
	9	0.23025851 (+01)	0.58648482 (-04)
	10	0.23025851 (+01)	0.29324216 (-04)
	11	0.23025851 (+01)	0.14662108 (-04)
	12	0.23025851 (+01)	0.73310541 (-05)
	13	0.23025851 (+01)	0.36655270 (-05)

Table III.4.1.

	i	x_i	
Halley	0	.4300000000000000 (01)	.2000000000000000 (01)
	1	.3336155282457216 (01)	.1035972419924183 (01)
	2	.2560818009367738 (01)	.2596797949731372 (00)
	3	.2308175634684460 (01)	.5683785304496196 (-02)
	4	.2302585151186738 (01)	.6120489087942105 (-07)
	5	.2302585092994046 (01)	-.3759322471455472 (-17)
Tangent Hyperbolas	0	.4300000000000000 (01)	.2000000000000000 (01)
	1	.3337356399057231 (01)	.1034771307502802 (01)
	2	.2561541506081360 (01)	.2589564130873139 (00)
	3	.2308222334300647 (01)	.5637241306601315 (-02)
	4	.2302585152707625 (01)	.5971357897526734 (-07)
	5	.2302585092994046 (01)	.1443269364993953 (-16)

Table III.4.2.

§ 5. INITIAL VALUE PROBLEMS

The successive approximations x_i in an iterative procedure will now be real-valued functions. Let $X = C'([0,T])$ and $Y = C([0,T])$ denote the set of all real-valued functions that are respectively continuously differentiable and continuous on the real interval $[0,T]$.

Consider the equation

$$\frac{dx}{dt} - f(t,x) = 0 \qquad x(0) = c \tag{III.5.1}$$

for $t \in [0,T]$.

We could restrict ourselves to the set $C_c'([0,T]) = \{x \in C'([0,T]) \mid x(0) = c\}$ and try to find a zero $x^*(t)$ of the following operator

$$F: C_c'([0,T]) \subset X \to C([0,T]): x \to \frac{dx}{dt} - f(t,x)$$

starting from an initial approximation $x_0(t)$ that satisfies $x_0(0) = c$, and computing corrections $(x_{i+1}-x_i)(t)$ that satisfy $(x_{i+1}-x_i)(0) = 0$.

We calculate the necessary derivatives:

$$F'(x_0) : C'([0,T]) \to C([0,T]) : x \to \left(\frac{d}{dt} - \frac{\partial f(t,x)}{\partial x}\Big|_{x=x_0(t)}\right) x$$

$$F''(x_0) : C'([0,T]) \times C'([0,T]) \to C([0,T]) : (x,x) \to \frac{\partial^2 f(t,x)}{\partial x^2}\Big|_{x = x_0(t)} . x^2$$

For the calculation of the Newton-correction $a_0(t)$ we have to solve the linear problem

$$F'(x_0)a_0 = -F(x_0) \tag{III.5.2}$$

and iterate

$$x_1(t) = x_0(t) + a_0(t) = x_0(t) - F'(x_0)^{-1}F(x_0)$$

One can prove that the solution of (III.5.2) is [41 pp. 170]

$$a_0(t) = -\int_0^t e^{A_0(s) - A_0(t)} F(x_0)(s)\, ds$$

where

$$A_0(t) = -\int_0^t \frac{\partial f(s,x(s))}{\partial x}\Big|_{x = x_0(s)} ds$$

The whole procedure can be repeated to calculate the next iterationsteps.

For the Chebyshev- or Halley-iteration one has to solve two linear problems:

$$F'(x_0)\, a_0 = - F(x_0) \qquad F'(x_0)\, b_0 = F''(x_0)\, a_0^2 \tag{III.5.3}$$

and iterate respectively

$$x_1(t) = x_o(t) + a_o(t) - \frac{1}{2}b_o(t)$$

or

$$x_1(t) = x_o(t) + \frac{a_o^2(t)}{a_o(t) + \frac{1}{2}b_o(t)}$$

We now turn to some examples.

Consider the nonlinear initial value problem

$$\frac{dx}{dt} - (1+x^2) = 0$$

$$x(0) = 0$$

for $t \in [0,T]$.

We will calculate $x_1(t)$ starting from $x_o(t) = t$ for the Newton-, Chebyshev- and Halley-iteration. Observe that:

$$A_o\ (t) = -t^2$$

$$F''\ (x_o)\ x^2 = -2x^2$$

$$-F\ (x_o) = t^2$$

$$a_o\ (t) = \int_0^t e^{t^2-s^2}\ s^2 ds = \frac{t^3}{3} + \frac{2t^5}{15} + \frac{4t^7}{105} + \frac{8t^9}{945} + \frac{16t^{11}}{10395} + \cdots$$

(term by term integration)

$$b_o\ (t) = -\int_0^t 2e^{t^2-s^2}[a_o(s)]^2\ ds = (-2)\ (\frac{t^7}{63} + \frac{38t^9}{2835} + \frac{992t^{11}}{155925} + \cdots)$$

The next iterationsteps are:

$$x_1\ (t) = t + \frac{1}{3}t^3 + \frac{2}{15}t^5 + \frac{4}{105}t^7 + \frac{8}{945}t^9 + \cdots \text{ (Newton)}$$

$$x_1\ (t) = t + \frac{1}{3}t^3 + \frac{2}{15}t^5 + \frac{17}{315}t^7 + \frac{62}{2835}t^9 + \frac{16}{2025}t^{11} + \cdots \text{ (Chebyshev)}$$

$$x_1\ (t) = t + \frac{1}{3}t^3 + \frac{2}{15}t^5 + \frac{17}{315}t^7 + \frac{62}{2835}t^9 - \frac{91369}{81860625}t^{11} + \cdots \text{ (Halley)}$$

For $T < \frac{\Pi}{2}$ the exact solution is

$$x^\star\ (t) = \text{tg}\ t = t + \frac{1}{3}t^3 + \frac{2}{15}t^5 + \frac{17}{315}t^7 + \frac{62}{2835}t^9 + \frac{4146}{467775}t^{11} + \cdots$$

Initial value problems correspond to Volterra integral equations. So equation (III.5.1) can be transformed into the following nonlinear integral equation:

$$F(x) = x(t) - c - \int_0^t f(s,x(s))\ ds$$

Now $F'(x_o) = I_x - V_o$ and $F''(x_o)\ x^2 = -\int_0^t \frac{\partial^2 f\ (s,x\ (s))}{\partial x^2}\Big|_{x = x_o(s)} x^2(s)\ ds$

where I_x: $x \to x$ is the identity operator and $V_o x = \int_0^t \frac{\partial f}{\partial x}(s,x(s))\big|_{x = x_o(s)} x(s)ds.$

So $F'(x_o)^{-1} = \sum_{n=0}^{\infty} V_o^n$ if $\|V_o\| < 1$. If we rewrite $F'(x_o)^{-1}x = (I_X + \sum_{n=1}^{\infty} V_o^n)\, x =$ $x + V_o\, (F'(x_o)^{-1}x)$ the equations (III.5.2) and (III.5.3) can be solved iteratively:

$$a_o^{(o)}(t) = 0$$

$$\begin{aligned} a_o^{(j)}(t) &= -F(x_o)(t) + V_o\, a_o^{(j-1)}(t) \\ &= -x_o(t) + c + \int_0^t f(s,x_o(s))ds + \int_0^t \frac{\partial f}{\partial x}(s,x(s))\Big|_{x = x_o(s)} a_o^{(j-1)}(s)\, ds \end{aligned}$$

$$b_o^{(o)}(t) = 0$$

$$\begin{aligned} b_o^{(j)}(t) &= F''(x_o)\, a_o^2(t) + V_o b_o^{(j-1)}(t) \\ &= -\int_0^t \frac{\partial^2 f\,(s,x(s))}{\partial x^2}\Big|_{x=x_o(s)} a_o^2\,(s)ds + \int_0^t \frac{\partial f}{\partial x}(s,x(s))\Big|_{x=x_o(s)} b_o^{(j-1)}(s)ds \end{aligned}$$

where $a_o\,(t)$ is the last approximation $a_o^{(j)}\,(t)$ for the Newton-correction.
For our example where $f(t,x) = 1 + x^2$ and $c=0$, we get the iterationsteps:

$$x_1\,(t) = t + \frac{1}{3}t^3 + \frac{2}{15}t^5 + \frac{4}{105}t^7 + \frac{8}{945}t^9 + \ldots \text{ (Newton)}$$

$$x_1\,(t) = t + \frac{1}{3}t^3 + \frac{2}{15}t^5 + \frac{17}{315}t^7 + \frac{62}{2835}t^9 + \frac{16}{2025}t^{11} + \ldots \text{ (Chebyshev)}$$

$$x_1\,(t) = t + \frac{1}{3}t^3 + \frac{2}{15}t^5 + \frac{17}{315}t^7 + \frac{62}{2835}t^9 - \frac{91369}{81860625}t^{11} + \ldots \text{ (Halley)}$$

Let us now turn to an example where the method of Halley, which is newly introduced here in (III.2.5), proves to be much better than the methods resulting from the Padé approximation problem of order (n,o) for G. Consider the equation

$$e^{x(t)}\frac{dx}{dt} - (0.1 + \varepsilon) = 0$$
$$x(1) = \ell n \varepsilon$$

for $t \in [1,T]$ with ε a small nonzero positive number and T large. We are looking for a zero $x^\star(t)$ of the nonlinear operator

$$F\colon x \rightarrow e^x\frac{dx}{dt} - (0.1 + \varepsilon) = y$$

The inverse operator

$$G\colon y \rightarrow \ell n\,(\varepsilon t + \int_1^t (0.1 + y)\, ds) = x$$

comes nearby a singularity for $y = -0.1$, thus in the neighbourhood of $y = 0$.
The exact solution is $x^\star(t) = \ell n\,(\varepsilon t + 0.1(t-1))$. Let us take our initial approximatior $x_o(t) = \ell n\, \varepsilon t$. The derivatives at x_o are

$$F'(x_0)\, x = e^{x_0(t)} \left(x \frac{dx_0}{dt} + \frac{dx}{dt}\right)$$

$$F''(x_0)\, x^2 = e^{x_0(t)}\, x(t) \left(2\frac{dx}{dt} + x\frac{dx_0}{dt}\right)$$

For $x_0(t) = \ell n\, \varepsilon t$:

$$F'(x_0)\, x = \varepsilon t \left(\frac{dx}{dt} + \frac{1}{t} x\right)$$

$$F''(x_0)\, x^2 = \varepsilon t.x.\left(2\frac{dx}{dt} + \frac{1}{t} x\right)$$

For the Newton-correction we have to solve the linear equation

$$\frac{da_0}{dt} + \frac{1}{t} a_0\, (t) = \frac{0.1}{\varepsilon t}$$

The solution is constructed in the same way as for (III.5.2):

$$a_0(t) = \int_1^t e^{A_0(s)-A_0(t)} \frac{0.1}{\varepsilon s}\, ds$$

where

$$A_0(t) = \int_1^t \frac{1}{s}\, ds = \ell n t$$

So

$$a_0(t) = \frac{0.1}{\varepsilon t}\, (t-1)$$

For the Chebyshev- and Halley-iteration we need the $b_0(t)$:

$$b_0(t) = \int_1^t \frac{s}{t} \left(\frac{0.1}{\varepsilon}\right)^2 \frac{s^2-1}{s^3}\, ds = [a_0(t)]^2$$

because

$$F''\,(x_0)\, a_0^2 = \left(\frac{0.1}{\varepsilon}\right)^2 \frac{t^2-1}{t^2}\, \varepsilon$$

The next iterationstep is:

$$x_1(t) = \ell n \varepsilon t + \frac{0.1}{\varepsilon} \frac{t-1}{t} \qquad \text{(Newton)}$$

$$x_1(t) = \ell n \varepsilon t + \frac{0.1}{\varepsilon} \frac{t-1}{t} \left(1 - \frac{0.1}{2\varepsilon} \frac{t-1}{t}\right) \qquad \text{(Chebyshev)}$$

$$x_1(t) = \ell n \varepsilon t + \frac{0.1}{\varepsilon} \frac{t-1}{t} \Big/ \left(1 + \frac{0.1}{2\varepsilon} \frac{t-1}{t}\right) \qquad \text{(Halley)}$$

$$x_1(t) = (\ell n \varepsilon t)^2 \Big/ \left(\ell n \varepsilon t - \frac{0.1}{\varepsilon} \frac{t-1}{t}\right) \qquad \text{(iteration (III.2.3))}$$

When we compare $\|x^\star(t) - x_1(t)\|_\infty = \sup_{t \in [1,T]} |x^\star(t) - x_1(t)|$ for the different procedures (see also figures III.5.1 - III.5.4 for the picture of the different functions $|x^\star(t) - x_1(t)|$) we see that for $\varepsilon = 0.01$ and T very large:

$$\|x^{\star}-x_1\|_{\infty} \simeq 10-\ln 11 \simeq 7.60 \qquad \text{(Newton)}$$

$$\|x^{\star}-x_1\|_{\infty} \simeq 40+\ln 11 \simeq 42.40 \qquad \text{(Chebyshev)}$$

$$\|x^{\star}-x_1\|_{\infty} \simeq -\frac{10}{6}+\ln 11 \simeq 0.73 \qquad \text{(Halley)}$$

$$\|x^{\star}-x_1\|_{\infty} \simeq 10-\ln 11 \simeq 7.60 \qquad \text{(iteration (III.2.3))}$$

Also the function-values for $t=2$ and $\varepsilon=0.01$ illustrate that the iterative procedures that take into account the singularity of the operator G in the neighbourhood of 0, are much more accurate:

$$x^{\star}(2) = -2.12026354$$

$$x_1(2) = 1.08797700 \qquad \text{(Newton)}$$

$$x_1(2) = -11.4120230 \qquad \text{(Chebyshev)}$$

$$x_1(2) = -2.48345158 \qquad \text{(Halley)}$$

$$x_1(2) = -1.71722223 \qquad \text{(iteration (III.2.3))}$$

Figure III.5.1.:

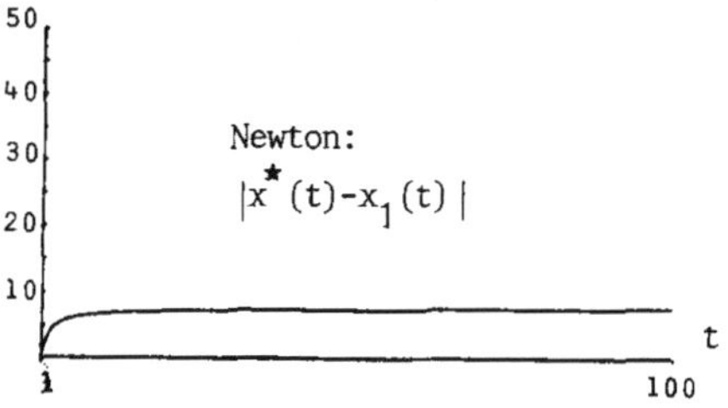

Figure III.5.2.:

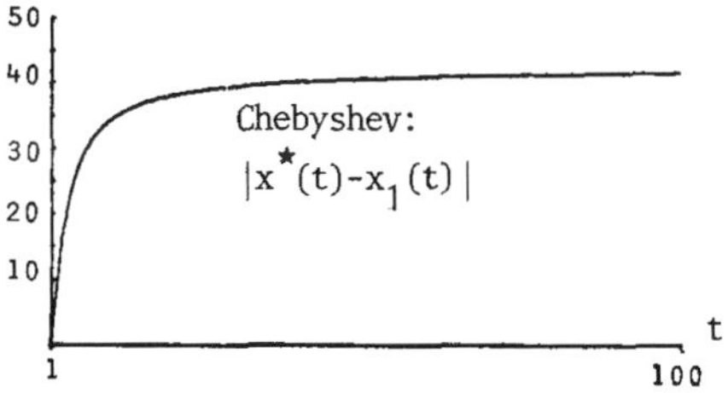

Figure III.5.3.:

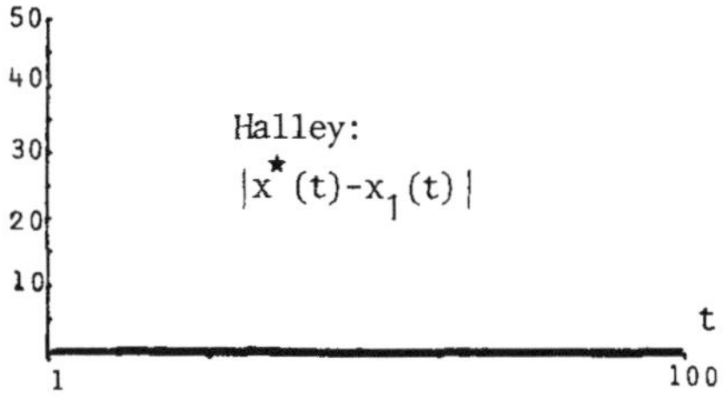

Figure III.5.4.:

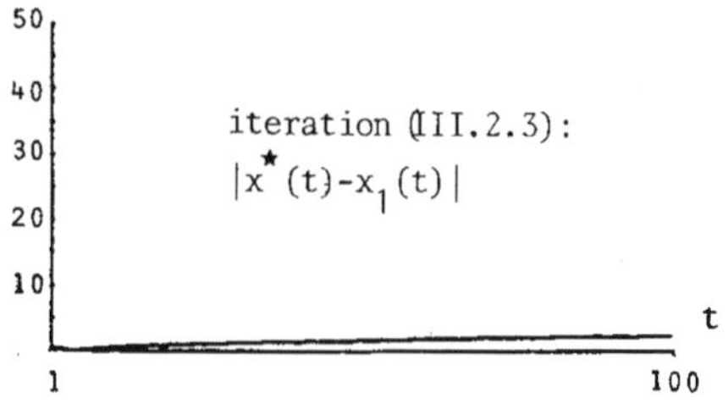

An iterative method resulting from the solution of the Padé approximation problem of order (n,m) for G with $m > o$, is also very useful when there are several singularities in the solution $x^{\star}(t)$ itself, because the rational approximations $x_i(t)$ can simulate certain singularities. We emphasize the fact that discontinuities cause difficulties when discretisation techniques are used. We will illustrate the advantage of the use of Halley's method and iteration (II.2.3) by an example.

Suppose we want to solve

$$F(x) = \frac{dx}{dt} + x^2 = 0$$

$$x(0) = -1$$

for $t \in [0,\frac{1}{2}] \cup [\frac{3}{2},T]$ with T large.

The solution $x^{\star}(t) = \frac{1}{t-1}$.

As an initial approximation we take $x_0(t) = -1$ and we calculate

$$F(x_0) = 1$$

$$F'(x_0)\,x = \frac{dx}{dt} - 2x$$

$$F''(x_0)\,x^2 = 2x^2$$

For the Newton-correction we have to solve the linear problem

$$\frac{d\,a_0}{dt} - 2a_0(t) = -1$$

The solution is constructed in the same way as previously

$$a_0(t) = -\int_0^t e^{A_0(s)-A_0(t)}\,ds$$

with

$$A_0(t) = -\int_0^t 2ds = -2t$$

So

$$a_0(t) = \frac{1}{2}\,(1-e^{2t})$$

Now we calculate the $b_0(t)$ for the Chebyshev- and Halley-iteration

$$\frac{d\,b_0}{dt} - 2b_0(t) = \frac{1}{2}(1-e^{2t})^2$$

So

$$b_0(t) = \int_0^t e^{A_0(s)-A_0(t)} \frac{1}{2}(1-e^{2s})^2 \, ds$$
$$= \frac{1}{4}(e^{4t}-1) - te^{2t}$$

The next iterationsteps are

$$x_1(t) = -\frac{1}{2}(1+e^{2t}) \qquad \text{(Newton)}$$

$$x_1(t) = -\frac{1}{2}e^{2t}(1-t) - \frac{1}{8}(e^{4t}+3) \qquad \text{(Chebyshev)}$$

$$x_1(t) = \frac{te^{2t} + \frac{1}{4}(e^{4t}-1)}{-(1+t)e^{2t} + \frac{1}{4}(e^{4t}+3)} \qquad \text{(Halley)}$$

$$x_1(t) = \frac{-2}{3 - e^{2t}} \qquad \text{(iteration (III.2.3))}$$

The exact solution $x^\star(t)$ has a pole in t=1. The iterationsteps $x_1(t)$, obtained by making use of the solution of the Padé approximation problem of order (1,1) and (0,1) are more accurate than the Newton- and Chebyshev-iterationsteps, because they approximate the pole of $x^\star(t)$ respectively by a pole in t=1.01993442 (Halley's method) and t=0.54930615 (iteration (III.2.3)). So they also approximate $x^\star(t)$ well beyond the discontinuity while for the Newton- and Chebyshev-iterationsteps $\lim_{t\to\infty} x_1(t) = -\infty$.

To illustrate this we compare the function-values for t=3/2:

$$x^\star(\tfrac{3}{2}) = 2.000$$

$$x_1(\tfrac{3}{2}) = -10.54 \qquad \text{(Newton)}$$

$$x_1(\tfrac{3}{2}) = -45.78 \qquad \text{(Chebyshev)}$$

$$x_1(\tfrac{3}{2}) = 2.544 \qquad \text{(Halley)}$$

$$x_1(\tfrac{3}{2}) = 0.117 \qquad \text{(iteration (III.2.3))}$$

§ 6. BOUNDARY VALUE PROBLEMS

Consider the equation

$$\frac{d^2x}{dt^2} - f(t,x) = 0$$

$$x(0) = 0 = x(1)$$

for $t \in [0,1]$.

Let $X = C''([0,1])$ denote the set of all real-valued functions that are twice continuously differentiable. Then we look for a zero of the operator

$$F: \{x \in C''([0,1]) \mid x(0)=0=x(1)\} \subset X \to C([0,1]) \;:\; x \to \frac{d^2x}{dt^2} - f(t,x)$$

The Newton-correction $a_0(t)$ is the solution of the following boundary value problem

$$\frac{d^2a_0}{dt^2} - \frac{\partial f}{\partial x}\Big|_{x=x_0(t)} \cdot a_0(t) = \frac{-d^2x_0}{dt^2} + f(t,x_0(t)) = v_0(t)$$

Since boundary value problems correspond to Fredholm integral equations, the Newton-correction is also the solution of the following linear Fredholm integral equation of the second kind

$$a_0(t) - \int_0^1 G(t,s)\,\frac{\partial f}{\partial x}(s,x(s))\Big|_{x=x_0} \cdot a_0(s)ds = \int_0^1 G(t,s)v_0(s)ds = w_0(t) \qquad \text{(III.6.1)}$$

where

$$G(t,s) = \begin{cases} s(t-1) & \text{for } 0 \le s \le t \\ t(s-1) & \text{for } t \le s \le 1 \end{cases} \qquad [41 \text{ pp. } 176]$$

This linear equation can be written as

$$(I_x - L)\; a_0(t) = w_0(t)$$

where

I_x: $x(t) \to x(t)$ is the identity operator

$$L\, a_0(t) = \int_0^1 L(t,s)\; a_0(s)\; ds$$

with

$$L(t,s) = \begin{cases} s(t-1)\,\frac{\partial f}{\partial x}(s,x(s))\Big|_{x=x_0} & \text{for } 0 \le s \le t \\ t(s-1)\,\frac{\partial f}{\partial x}(s,x(s))\Big|_{x=x_0} & \text{for } t \le s \le 1 \end{cases}$$

If this linear operator (I_x-L) is bounded then $(I_x-L)^{-1}$ exists if and only if a linear bounded operator K with inverse K^{-1} exists such that $\|I_x-K(I_x-L)\| < 1$.
Then $(I_x-L)^{-1} = \sum_{n=0}^{\infty} [I_x-K(I_x-L)]^n K$ [41 pp. 43]. Let us take $K = I_x$ here. Then $I_x-K(I_x-L) = L$.

$$\text{Now } \|L\| = \sup_{\|x\|=1} \|Lx\| \le \max_{[0,1]} \int_0^1 |\, L(t,s)\,|\; ds$$

$$\le \left\|\frac{\partial f}{\partial x}\Big|_{x=x_0(t)}\right\| \cdot \max_{[0,1]} \int_0^1 |\, G(t,s)\,|\; ds$$

$$= \frac{1}{8}\left\|\frac{\partial f}{\partial x}\Big|_{x=x_0(t)}\right\|$$

where $\|\;\| = \max_{[0,1]} |\;|$.

So if $\|\frac{\partial f}{\partial x}\mid_{x=x_o}\|$ is small enough then $(I_X-L)^{-1} = \sum_{n=o}^{\infty} L^n$.

Again the Newton-correction can be computed iteratively

$$a_o^{(o)}(t) = 0$$

$$a_o^{(j)}(t) = w_o(t) + \int_0^1 L(t,s)\, a_o^{(j-1)}(s)\, ds$$

where

$$\|a_o(t) - a_o^{(j)}(t)\| \le \frac{\|L\|^{j+1}\ \|w_o\|}{1 - \|L\|}$$

The correction $b_o(t)$ can be calculated analogously, and the whole procedure can be repeated for the next iterationsteps. As an example we will solve the equation

$$\frac{d^2x}{dt^2} - (t\,x^2 - 1) = 0$$

$$x(0) = 0 = x(1)$$

for $t \in [0,1]$.

Let us take $x_o(t) = 0$. For this $f(t,x)$, $\|L\| = 1/54 < 1$.

The solution of equation (III.6.1) is

$$a_o(t) = \frac{1}{2}t(1-t)$$

The correction $b_o(t)$ is the solution of the boundary value problem

$$\frac{d^2b_o}{dt^2} - \frac{\partial f}{\partial x}\mid_{x=x_o(t)} \cdot b_o(t) = F''(x_o)\, a_o^2(t) = -2ta_o^2(t)$$

or converted into an integral equation

$$b_o(t) - \int_0^1 G(t,s)\, \frac{\partial f}{\partial x}(s,x(s))\mid_{x=x_o} \cdot b_o(s)ds = -\int_0^1 G(t,s)\frac{s^3}{2}(1-s)^2\, ds$$

So $b_o(t) = -\frac{1}{2}\left(\frac{t^7}{42} - \frac{t^6}{15} + \frac{t^5}{20} - \frac{t}{140}\right) = a_o(t)\left(\frac{t^5}{42} - \frac{3}{70}t^4 + \frac{1}{140}(t^3 + t^2 + t + 1)\right)$

The next iterationstep is

$$x_1(t) = \frac{1}{2}t\ (1-t) \qquad \text{(Newton)}$$

$$x_1(t) = \frac{1}{4}t\ (1-t)\left[2 - \frac{t^5}{42} + \frac{3}{70}t^4 - \frac{1}{140}(t^3 + t^2 + t + 1)\right] \qquad \text{(Chebyshev)}$$

$$x_1(t) = \frac{t(t-1)}{-2 - \frac{t^5}{42} + \frac{3}{70}t^4 - \frac{1}{140}(t^3 + t^2 + t + 1)} \qquad \text{(Halley)}$$

If we calculate $a_1(t)$ iteratively, we get

$$a_1^{(o)}(t) = 0$$

$$a_1^{(1)}(t) = \frac{1}{4}\left(\frac{t^7}{42} - \frac{t^6}{15} + \frac{t^5}{20} - \frac{t}{140}\right)$$

and for $x_2(t)$ in the Newton iteration

$$x_2(t) = x_1(t) + a_1^{(1)}(t)$$

$$= a_o(t) - \frac{1}{2} a_o(t)\left(\frac{t^5}{42} - \frac{3}{70}t^4 + \frac{1}{140}(t^3 + t^2 + t + 1)\right)$$

which is precisely one Chebyshev-iterationstep.

The solution of the boundary value problem has been calculated for discrete values $t_i = \frac{i}{200}$ $(i=0,\ldots,200)$ in the interval $[0,1]$, by means of subroutine DDØ2AD of the Harwell-library (based on a finite difference approximation to a linearized form of the equation) and also with the initial values $x_i = x(t_i) = 0$. After interpolation through the (t_i, x_i) we get the following picture of the solution $x^*(t)$

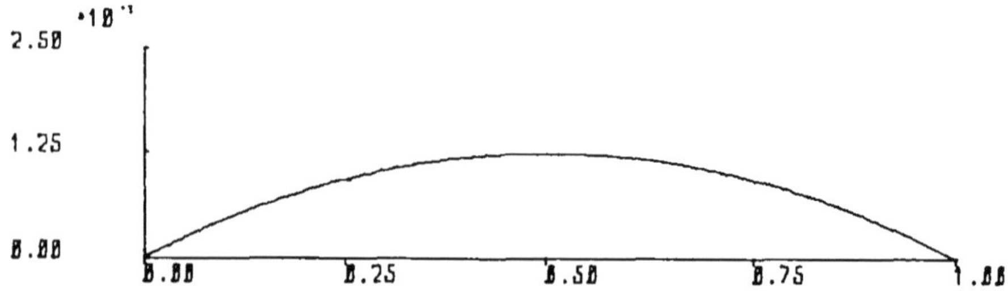

Figure III.6.1.

The different functions $x_1(t)$ mentioned above, give the same plot. We can also compare the function-values in some points (7 significant figures):

t	DDØ2AD	Newton	Chebyshev	Halley
0.25	0.0933169	0.0937500	0.0933121	0.0933141
0.50	0.1242918	0.1250000	0.1242839	0.1242879
0.75	0.0932114	0.0937500	0.0932053	0.0932084

Table III.6.1.

The functions $|x^*(t) - x_1(t)|$ for the different iterative schemes give the following plots:

Figure III.6.2.

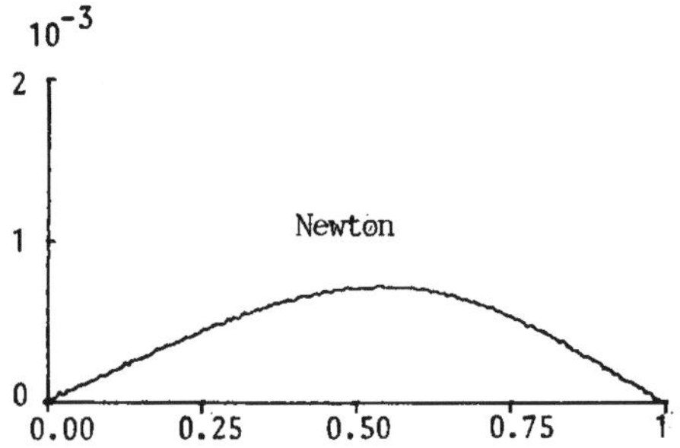

Figure III.6.3.

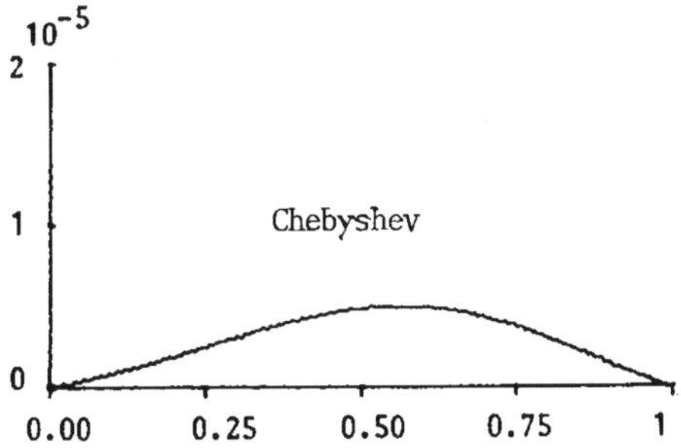

Figure III.6.4.

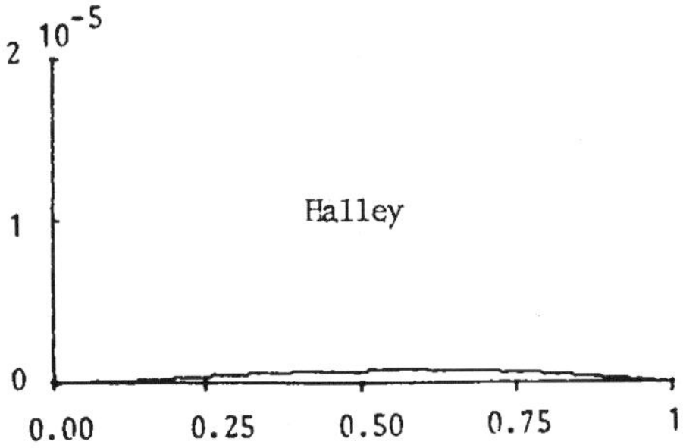

§ 7. PARTIAL DIFFERENTIAL EQUATIONS

Consider the following nonlinear equation which is of interest in gas dynamics

$$\Delta x(s,t) = \frac{\partial^2 x}{\partial s^2} + \frac{\partial^2 x}{\partial t^2} = x^2(s,t) \text{ for } (s,t) \text{ in } \Omega \subset \mathbb{R}^2$$

$$x(s,t) = r(s,t) \text{ on the boundary of the region } \Omega$$

A solution $x(s,t)$ is sought in the interior of Ω.

If $F(x) = \Delta x - x^2$, then

$$F'(x_o)\, x = \Delta x - 2x_o . x$$

$$F''(x_o)\, x^2 = -2x^2$$

The Newton-correction satisfies

$$\Delta a_o(s,t) - 2\, a_o(s,t) \cdot x_o(s,t) = x_o^2\,(s,t) - \Delta x_o(s,t) \tag{III.7.1}$$

$$a_o(s,t) = 0 \text{ on the boundary of the region } \Omega$$

Pohozaev has proved that [40]

$$\Delta x = x^2$$

$$x(s,t) = r(s,t) > 0 \text{ on the boundary of } \Omega$$

has a unique positive solution $x^*(s,t)$ and that the Newton iteration converges if the initial approximation x_o is the solution of the Laplace equation with the same Dirichlet boundary conditions:

$$\Delta x_o = 0$$

$$x_o(s,t) = r(s,t) > 0 \text{ on the boundary of } \Omega$$

This initial approximation cancels the term $-\Delta x_o$ in (III.7.1). Instead of solving (III.7.1) we can again rewrite it as a linear integral equation of Fredholm type and second kind by means of the Green's function $K(s,t,u,v)$ for Ω:

$$a_o(s,t) = 2 \iint_\Omega K(s,t,u,v)\, a_o(u,v)\, x_o(u,v)\, du\, dv + \iint_\Omega K(s,t,u,v)\, x_o^2(u,v)\, du\, dv \tag{III.7.2}$$

If $\Omega = [0,1] \times [0,1]$ then

$$K(s,t,u,v) = \frac{-4}{\Pi^2} \sum_{\substack{j=1\\k=1}}^{\infty} \frac{\sin k\pi s \sin j\pi t \sin k\pi u \sin j\pi v}{j^2 + k^2}$$

$$\approx \frac{-4}{\Pi^2} \sum_{\substack{j=1\\k=1}}^{n} \frac{\sin k\pi s \sin j\pi t \sin k\pi u \sin j\pi v}{j^2 + k^2}$$

For $r(s,t) = 1$ the initial approximation $x_o(s,t) = 1$. We compute $a_o(s,t)$ by repeated substitution in (III.7.2), where we use the indicated approximation for $K(s,t,u,v)$:

$$a_o^{(o)}(s,t) = 0$$

$$a_o^{(1)}(s,t) = -\frac{16}{\Pi^4} \sum_{\substack{j=1\\k=1\\j \text{ odd}\\k \text{ odd}}}^{n} \frac{\sin k\pi s \sin j\pi t}{(k^2 + j^2)\, kj}$$

We will denote $\sum_{\substack{j,k=1 \\ j,k \text{ odd}}}^{n}$ from now on by $\sum_{j,k=1}^{n''}$.

The function $b_o(s,t)$ is the solution of

$$b_o(s,t) = 2 \iint_\Omega K(s,t,u,v)\ b_o(u,v)\ x_o(u,v)\ du\ dv - 2 \iint_\Omega K(s,t,u,v)\ a_o^2(u,v)\ du\ dv$$

since $F''(x_o)\ a_o^2 = -2a_o^2$.

So for $r(s,t) = 1$ and $\Omega = [0,1] \times [0,1]$ we get

$$b_o^{(o)}(s,t) = 0$$

$$b_o^{(1)}(s,t) = \frac{2^{15}}{\pi^{12}} \sum_{\substack{j,k=1 \\ \ell,m=1 \\ i,h=1}}^{n''} \frac{ih \sin i\pi s \sin h\pi t}{(i^2+h^2)(j^2+k^2)(\ell^2+m^2)\ P(i,k,\ell)\ P(h,j,m)}$$

where

$$P(i,k,\ell) = (i-k+\ell)(i+k-\ell)(i-k-\ell)(i+k+\ell)$$

Greenspan has proved that the solutions of the following finite systems which are the result of a discretisation of (III.7.1), converge to the solution of $\Delta x = x^2$ with the given Dirichlet boundary conditions, as the mesh size h approaches zero [26]:

let $x_{ij} = x(s_i,t_j) = x(ih,jh)$

construct $x_{ij}^{(k)}$ in terms of $x_{ij}^{(k-1)}$ as follows

$$x_{i,o}^{(k)} = x_{o,j}^{(k)} = x_{i,m}^{(k)} = x_{m,j}^{(k)} = 1 \text{ for } i,j = o,\dots,m \text{ and } h = \frac{1}{m} \qquad \text{(III.7.3)}$$

$$x_{ij}^{(o)} = 1$$

$$x_{ij}^{(k)}\left(-2x_{ij}^{(k-1)} - \frac{4}{h^2}\right) + \frac{1}{h^2}\left(x_{i+1,j}^{(k)} + x_{i-1,j}^{(k)} + x_{i,j+1}^{(k)} + x_{i,j-1}^{(k)}\right) = -[x_{ij}^{(k-1)}]^2$$

The procedure terminates when $\max_{i,j} |x_{ij}^{(k)} - x_{ij}^{(k-1)}| \le \varepsilon$ and this final $x_{ij}^{(k)}$ is defined to be the solution. We shall now compare the function-values of the different iteration-steps $x_1(s,t)$ (Newton, Chebyshev, Halley) and the solution of (III.7.3) for h=1/100 and ε=5.0(-9). For the calculation of K(s,t,u,v) we have taken n=5. The functions $x_1(s,t)$ all give the plot drawn in figure III.7.1.

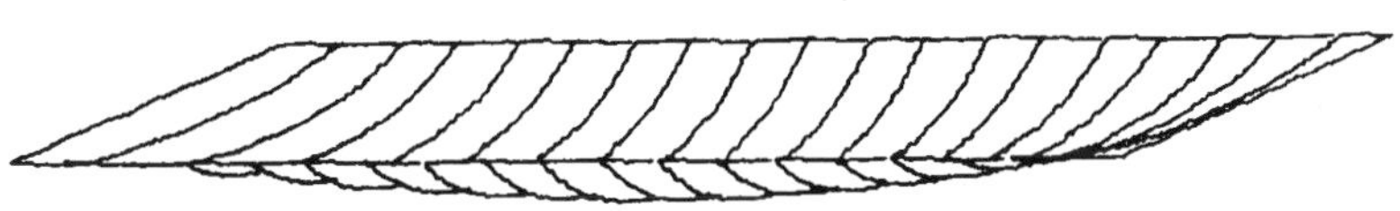

Figure III.7.1.

	t \ s	0.25	0.50	0.75
Newton	0.25	0.954473792	0.942281229	0.954473792
	0.50	0.942281229	0.925794323	0.942281229
	0.75	0.954473792	0.942281229	0.954473792
Chebyshev	0.25	0.954360724	0.942115745	0.954360724
	0.50	0.942115745	0.925550785	0.942115745
	0.75	0.954360724	0.942115269	0.954360724
Halley	0.25	0.954360443	0.942115269	0.954360443
	0.50	0.942115269	0.925549983	0.942115269
	0.75	0.954360443	0.942115269	0.954360443
h = 1/100	0.25	0.958513709	0.947882237	0.958513709
	0.50	0.947882192	0.933717325	0.947882192
	0.75	0.958513647	0.947882149	0.958513647

Table III.7.1.

§ 8. NONLINEAR INTEGRAL EQUATION OF FREDHOLM TYPE

A general nonlinear Fredholm integral equation may be written in the form

$$F(x) = \int_a^b K(t,s,x(t),x(s))\, ds = 0 \text{ for } a \le t \le b$$

We will treat the equation

$$F(x) = x(t) - 1 - \frac{\lambda}{2} x(t) \int_0^1 \frac{t}{t+s} x(s)\, ds = 0 \qquad \text{(III.8.}$$

$$\text{for } 0 \le t \le 1 \text{ and } 0 \le \lambda \le 1$$

which was derived by Chandrasekhar [10]

If we write

$$Lx = \int_0^1 \frac{t}{t+s} x(s)\, ds$$

then

$$F'(x_0)\, x = x - \frac{\lambda}{2}(x.Lx_0 + x_0.Lx)$$

$$F''(x_0)\, x^2 = -\lambda x.Lx$$

For $x_0 = 1$ the Newton-correction is found by solving

$$(1 - \frac{\lambda}{2}t\ \ell n\ \frac{t+1}{t})\ a_o(t) - \frac{\lambda}{2}\int_0^1 \frac{t}{t+s}\ a_o(s)\ ds = \frac{\lambda}{2}\ t\ \ell n\ \frac{t+1}{t}$$

which can be converted in a linear integral equation of Fredholm type and second kind

$$a_o(t) - \frac{\frac{\lambda}{2}}{1 - \frac{\lambda}{2}\ t\ \ell n\ \frac{t+1}{t}} \int_0^1 \frac{t}{t+s}\ a_o(s)ds = \frac{\frac{\lambda}{2}\ t\ \ell n\ \frac{t+1}{t}}{1 - \frac{\lambda}{2}\ t\ \ell n\ \frac{t+1}{t}}$$

The equation can be written in the form

$$(I_X - \mathcal{L})\ a_o(t) = \frac{\frac{\lambda}{2}\ t\ \ell n\ \frac{t+1}{t}}{1 - \frac{\lambda}{2}\ t\ \ell n\ \frac{t+1}{t}}$$

where

$$\mathcal{L}x = \int_0^1 \frac{\frac{\lambda}{2}\ t}{(1 - \frac{\lambda}{2}\ t\ \ell n\ \frac{t+1}{t})(t+s)}\ x(s)\ ds$$

Now $\|I_X - \mathcal{L}\| \le 1 + \frac{\frac{\lambda}{2}\ell n2}{1-\frac{\lambda}{2}\ell n2} < 2$ where $\|\ \| = \max\limits_{[0,1]} |\ |$ and so we can try to invert $I_X - \mathcal{L}$ as we previously did (cfr. boundary value problems). Take again $K = I_X$. Then $I_X - K(I_X - \mathcal{L}) = \mathcal{L}$ with $\|\mathcal{L}\| \le \frac{\lambda\ell n2}{2-\lambda\ell n2} < 1$ and $(I_X - \mathcal{L})^{-1} = \sum\limits_{n=o}^{\infty} \mathcal{L}^n$.

The Newton-correction can be computed as follows

$$a_o^{(o)}\ (t) = 0$$

$$a_o^{(j)}\ (t) = \frac{\frac{\lambda}{2}\ t\ \ell n\ \frac{t+1}{t}}{1 - \frac{\lambda}{2}\ t\ \ell n\ \frac{t+1}{t}} + \mathcal{L}a_o^{(j-1)}\ (t) \qquad \text{(III.8.2)}$$

The correction $b_o\ (t)$ is calculated analogously

$$b_o^{(o)}\ (t) = 0$$

$$b_o^{(j)}\ (t) = \frac{-\ \lambda\ a_o\ (t)}{1 - \frac{\lambda}{2}\ t\ \ell n\ \frac{t+1}{t}} \int_0^1 \frac{t}{t+s}\ a_o(s)\ ds + \mathcal{L}b_o^{(j-1)}\ (t) \qquad \text{(III.8.3)}$$

where $a_o(t)$ is the last approximation $a_o^{(j)}(t)$ for the Newton-correction.

It has been proved [47] that the exact solution of (III.8.1) is

$$x^\star\ (t) = \exp\left(\frac{-t}{\pi} \int_0^{\frac{\pi}{2}} \frac{\ell n\ (1 - \lambda\theta\ \text{cotg}\ \theta)}{t^2\ \sin^2\theta + \cos^2\theta}\ d\theta\right) \quad 0 \le \lambda \le 1 \qquad \text{(III.8.4)}$$

if we take j=1 for the calculation of $a_o(t)$ and $b_o(t)$ we get

$$x_1(t) = 1 + \frac{\frac{\lambda}{2} t \ln \frac{t+1}{t}}{1 - \frac{\lambda}{2} t \ln \frac{t+1}{t}} = 1 + a_o^{(1)}(t) \qquad \text{(Newton)}$$

$$x_1(t) = 1 + a_o^{(1)}(t) + \frac{\frac{\lambda}{2} a_o^{(1)}(t)}{1 - \frac{\lambda}{2} t \ln \frac{t+1}{t}} \int_0^1 \frac{t}{t+s} a_o^{(1)}(s)\, ds \qquad \text{(Chebyshev)}$$

$$x_1(t) = 1 + \frac{a_o^{(1)}(t)}{1 - \frac{\frac{\lambda}{2}}{1 - \frac{\lambda}{2} t \ln \frac{t+1}{t}} \int_0^1 \frac{t}{t+s} a_o^{(1)}(s)\, ds} \qquad \text{(Halley)}$$

Rall mentions the fact that $x_o(t) = 1$ is a satisfactory initial approximation for the Newton-iteration only if [41 pp. 77]

$$0 \le \lambda \le \frac{\sqrt{2}-1}{\ln 2} = 0.59758\ldots$$

For other λ we need other initial approximations. If we want to know the solutions $x^*(t)$ for $\lambda = \frac{\ell}{10}$ (ℓ=0,...,10) we could use a tactic known as continuation: the solution for $\lambda = \frac{\ell}{10}$ is used as an initial approximation for the calculation of the solution for $\lambda = \frac{\ell+1}{10}$. Now for $\lambda = 0$ the exact solution of (III.8.1) is $x^*(t) = 1$.
For the computation of the integrals in (III.8.2) and (III.8.3) we have used the nine-point Gaussian integration rule [1 pp. 916]

$$\int_0^1 f(t)\, dt \cong \sum_{k=1}^{9} w_k f(t_k)$$

where

	t_k
k = 1	0.0159198802461869
2	0.0819844463366821
3	0.1933142836497048
4	0.3378732882980955
5	0.5000000000000000
6	0.6621267117019045
7	0.8066857163502952
8	0.9180155536633179
9	0.9840801197538131

and the w_k are the solution of the linear system

$$\sum_{k=1}^{9} t_k^{\ell-1} w_k = \frac{1}{\ell} \quad (\ell = 1,\ldots,9).$$

This integration rule enables us to calculate $a_o^{(j)}(t_k)$ and $b_o^{(j)}(t_k)$ to the desired accuracy. It also enables us to calculate further iterationsteps $x_{i+1}(t_k)$:

$$Lx_i(t_k) = t_k \sum_{\ell=1}^{9} \frac{w_\ell}{t_k + t_\ell} x_i(t_\ell) \quad k = 1,\ldots,9$$

$$F(x_i)(t_k) = -1 + x_i(t_k)\left(1 - \frac{\lambda}{2} Lx_i(t_k)\right)$$

$a_i^{(0)}\ (t_k) = 0$ and $b_i^{(0)}\ (t_k) = 0$

$$a_i^{(j)}\ (t_k) = \frac{1 - x_i\ (t_k)\ (1 - \frac{\lambda}{2} L\ x_i\ (t_k))}{1 - \frac{\lambda}{2} L\ x_i(t_k)} + \frac{\frac{\lambda}{2} x_i\ (t_k)\ t_k}{1 - \frac{\lambda}{2} L\ x_i(t_k)} \left(\sum_{\ell=1}^{9} \frac{w_\ell}{t_k+t_\ell} a_i^{(j-1)}\ (t_\ell)\right)$$

to the desired accuracy, and

$$b_i^{(j)}\ (t_k) = \frac{\frac{\lambda}{2}\ t_k}{1 - \frac{\lambda}{2} L\ x_i\ (t_k)} \left[\sum_{\ell=1}^{9} \frac{w_\ell}{t_k+t_\ell} (-2a_i(t_k)a_i(t_\ell) + x_i(t_k)b_i^{(j-1)}(t_\ell))\right]$$

to the desired accuracy, where $a_i(t_k)$ is the last approximation $a_i^{(j)}(t_k)$ to the Newton-correction. We can continue the iteration until

$$\max_{k=1,\dots,9} |x_i(t_k)-x_{i-1}(t_k)| \leq \varepsilon$$

We give the solution $x^\star(t_k)$ for $\lambda = \frac{\ell}{10}$ ($\ell=1,\dots,10$) and the number of iterationsteps needed in the different iterative procedures to achieve convergence to 8 decimal digits ($\varepsilon = 5.(-9)$). In [41 pp. 78] Rall has approximated the integral equation (III.8.1) by

$$\xi_k - 1 - \frac{\lambda}{2} \xi_k t_k \sum_{\ell=1}^{9} \frac{w_\ell\ \xi_\ell}{t_k+t_\ell} \text{ where } \xi_k = x\ (t_k) \text{ for } k = 1,\dots,9$$

This fixed point problem can be solved by repeated substitution and the method of continuation. The number of iterations required now to obtain convergence to eight decimal places is also shown in the following table. We notice a significant difference. All the computations are performed in double precision accuracy (about 16 decimal digits). For the calculation of $x^\star(t_k)$ for a chosen λ, (III.8.4) has been rewritten as follows [47] to remove singularities in the integrand for small t and great λ

$$x^\star\ (t) = \exp\ (z^\star\ (t))$$

with

$$z^\star\ (t) = \frac{1}{\pi} \int_0^{\frac{\pi}{2}} [f(\theta) - g(\theta) + h(\theta)]\ d\theta + z_2(t) - z_3(t).$$

$$f(\theta) = \lambda \text{ Arctg } (t \text{ tg } \theta) \frac{\theta \text{ cosec}^2\ \theta - \text{cotg } \theta}{1 - \lambda\ \theta \text{ cotg } \theta}$$

$$g(\theta) = \frac{\pi}{2} \lambda \text{ Arctg } (t \text{ tg } \theta)$$

$$h(\theta) = \frac{2\ t\ (1-\lambda)}{1 - \lambda + \frac{1}{3} \lambda\ \theta^2}$$

$$z_2(t) = \begin{cases} \frac{1}{2} \lambda \left\{\frac{\pi^2}{8} - \sum_{n=0}^{\infty} \frac{1}{(2n+1)^2} \left(\frac{1-t}{1+t}\right)^{2n+1}\right\} & \text{for } 1 \geq t > \sqrt{2} - 1 \\ \frac{1}{2} \lambda \left\{ \frac{1}{2} \ln t \ln \frac{1-t}{1+t} + \sum_{n=0}^{\infty} \frac{t^{2n+1}}{(2n+1)^2}\right\} & \text{for } 0 \leq t \leq \sqrt{2} - 1 \end{cases}$$

$$z_3(t) = \frac{2t}{\pi} \sqrt{\frac{3(1-\lambda)}{\lambda}} \text{ Arctg } \left(\frac{\pi}{2} \sqrt{\frac{\lambda}{3(1-\lambda)}}\right)$$

$x^*(t_k)$

NE = Newton
CH = Chebyshev
HA = Halley
FP = Fixed point

k	λ = 0.1	λ = 0.2	λ = 0.3	λ = 0.4	λ = 0.5	λ = 0.6	λ = 0.7	λ = 0.8	λ = 0.9	λ = 1.0
1	1.00336988	1.00687491	1.01053648	1.01438330	1.01845565	1.02281372	1.02755593	1.03286827	1.03921408	1.05118792
2	1.01089751	1.02250589	1.03495048	1.04840010	1.06309273	1.07938509	1.09786206	1.11963041	1.14744810	1.20857560
3	1.01829895	1.03817912	1.05995679	1.08406415	1.11111907	1.14206127	1.17845395	1.22329757	1.28417328	1.43721371
4	1.02435468	1.05121660	1.08114154	1.11490155	1.15362003	1.19903811	1.25411164	1.32464119	1.42564702	1.71373669
5	1.02892234	1.06117663	1.09755911	1.13919206	1.18773513	1.24580716	1.31794506	1.41326257	1.55603380	2.01277877
6	1.03220523	1.06840231	1.10959650	1.15722111	1.21342320	1.28164104	1.36793401	1.48472695	1.66599369	2.30601170
7	1.03445865	1.07339474	1.11797620	1.16988217	1.23165182	1.30739606	1.40445447	1.53811996	1.75110922	2.56462107
8	1.03589121	1.07658239	1.12335356	1.17805495	1.24350201	1.32428633	1.42867803	1.57409995	1.80995660	2.76255967
9	1.03664375	1.07826123	1.12619412	1.18238740	1.24981062	1.33332581	1.44173217	1.59367978	1.84249994	2.87963509
NE	3	3	3	4	4	4	4	4	5	17
CH	3	3	3	3	3	3	3	3	4	13
HA	3	3	3	3	3	3	3	3	3	12
FP	7	9	10	12	15	18	22	29	45	44293

Table III.8.1.

Table III.8.2.

$|x^{*}(t_k) - x_I(t_k)|$ $(\varepsilon = 5.(-9))$

k	λ = 0.1	λ = 0.2	λ = 0.3	λ = 0.4	λ = 0.5	λ = 0.6	λ = 0.7	λ = 0.8	λ = 0.9	λ = 1.0
1	3.7(-5)	7.5(-5)	1.1(-4)	1.5(-4)	1.9(-4)	2.3(-4)	2.7(-4)	3.0(-4)	3.5(-4)	3.9(-4)
2	5.0(-7)	9.4(-7)	1.3(-6)	1.6(-6)	1.9(-6)	2.0(-6)	2.1(-6)	2.1(-6)	2.1(-6)	7.1(-6)
3	8.9(-9)	4.7(-8)	1.2(-7)	2.2(-7)	3.6(-7)	5.4(-7)	7.7(-7)	1.0(-6)	1.4(-6)	1.2(-5)
4	7.8(-9)	3.2(-8)	7.6(-8)	1.4(-7)	2.2(-7)	3.3(-7)	4.8(-7)	6.6(-7)	9.0(-7)	2.8(-5)
5	$\le \varepsilon$	2.2(-8)	5.1(-8)	9.5(-8)	1.5(-7)	2.3(-7)	3.4(-7)	4.7(-7)	6.5(-7)	5.1(-5)
6	$\le \varepsilon$	1.7(-8)	3.9(-8)	7.3(-8)	1.2(-7)	1.8(-7)	2.6(-7)	3.7(-7)	5.3(-7)	7.8(-5)
7	$\le \varepsilon$	1.4(-8)	3.2(-8)	6.0(-8)	9.9(-8)	1.5(-7)	2.2(-7)	3.2(-7)	4.5(-7)	1.1(-4)
8	$\le \varepsilon$	1.2(-8)	2.9(-8)	5.3(-8)	8.8(-8)	1.3(-7)	2.0(-7)	2.8(-7)	4.1(-7)	1.3(-4)
9	$\le \varepsilon$	1.1(-8)	2.7(-8)	5.0(-8)	8.2(-8)	1.3(-7)	1.9(-7)	2.7(-7)	3.9(-7)	1.5(-4)

Table III.8.3.

$|x^{*}(t_k) - x_F(t_k)|$ $(\varepsilon = 5.(-9))$

k	λ = 0.1	λ = 0.2	λ = 0.3	λ = 0.4	λ = 0.5	λ = 0.6	λ = 0.7	λ = 0.8	λ = 0.9	λ = 1.0
1	3.7(-5)	7.5(-5)	1.1(-4)	1.5(-4)	1.9(-4)	2.3(-4)	2.7(-4)	3.1(-4)	3.5(-4)	3.9(-4)
2	5.0(-7)	9.4(-7)	1.3(-6)	1.6(-6)	1.9(-6)	2.0(-6)	2.1(-6)	2.1(-6)	2.1(-6)	9.7(-6)
3	8.8(-9)	4.7(-8)	1.2(-7)	2.2(-7)	3.6(-7)	5.4(-7)	7.7(-7)	1.0(-6)	1.4(-6)	2.0(-5)
4	7.8(-9)	3.2(-8)	7.5(-8)	1.4(-7)	2.2(-7)	3.3(-7)	4.8(-7)	6.6(-7)	8.9(-7)	4.4(-5)
5	5.3(-9)	2.2(-8)	5.1(-8)	9.4(-8)	1.5(-7)	2.3(-7)	3.3(-7)	4.7(-7)	6.5(-7)	7.8(-5)
6	$\le \varepsilon$	1.7(-8)	3.8(-8)	7.2(-8)	1.2(-7)	1.8(-7)	2.6(-7)	3.7(-7)	5.2(-7)	1.2(-4)
7	$\le \varepsilon$	1.4(-8)	3.2(-8)	5.9(-8)	9.8(-8)	1.5(-7)	2.2(-7)	3.1(-7)	4.5(-7)	1.6(-4)
8	$\le \varepsilon$	1.2(-8)	2.8(-8)	5.2(-8)	8.7(-8)	1.3(-7)	1.9(-7)	2.8(-7)	4.0(-7)	2.0(-4)
9	$\le \varepsilon$	1.1(-8)	2.6(-8)	4.8(-8)	8.1(-8)	1.2(-7)	1.8(-7)	2.6(-7)	3.8(-7)	2.2(-4)

The convergence to eight decimal places of the different methods of approximation does not imply that those eight digits are significant digits for $x^{\star}(t_k)$. For small t_k and great λ the iterative methods do not converge to $x^{\star}(t_k)$ but to a function in the neighbourhood of $x^{\star}(t_k)$. Let us denote by $x_I(t_k)$ the solution obtained by performing one of the iterative procedures Newton, Chebyshev or Halley (for each of the iterative procedures after a different number of iterationsteps) and let us denote by $x_F(t_k)$ the solution obtained after rewriting (III.8.1) as a fixed point problem.
In the tables III.8.2 and III.8.3 one can find $|x^{\star}(t_k) - x_I(t_k)|$ and $|x^{\star}(t_k) - x_F(t_k)|$ for k=1,...,9 and λ =0.1,...,1.0 . For small t_k (k=1,2) generally

$$x_F(t_k) \leq x^{\star}(t_k) \leq x_I(t_k)$$

Only for λ = 1.0 one notices slight differences.

§ 9. NUMERICAL STABILITY OF THE HALLEY-ITERATION FOR THE SOLUTION OF A SYSTEM OF NONLINEAR EQUATIONS

9.1. *Numerical stability of iterations*

Consider the numerical solution of the equation

$$F(x) = 0 \tag{III.9.1}$$

with $F: \mathbb{R}^p \to \mathbb{R}^p: x \to F(x)$, abstract analytic in 0 and assume that (III.9.1) has a simple root $x^{\star}$. We briefly repeat the definition of condition-number given by [48] Woźniakowski.
The consition-number should measure the sensitivity of the solution (output) with respect to changes in the data (input). We assume that F depends parametrically on a vector $d \in \mathbb{R}^q$, called data vector

$$F(x) = F(x;d)$$

Instead of the exact value F(x;d) we only have the computed value $f\ell(F(x;d))$ in t digit floating-point binary arithmetic. At best we can expect that $f\ell(F(x;d))$ is the exact value of a slightly perturbed operator at slightly perturbed data

$$f\ell(F(x;d)) = (I_x + \Delta F)\ F(x+\Delta x;d+\Delta d) \tag{III.9.2}$$

where I_x is the p x p unit-matrix, ΔF is a p x p matrix and

$$\begin{aligned} \|\Delta x\| &\leq C_1 \rho \|x\| \\ \|\Delta d\| &\leq C_2 \rho \|d\| \\ \|\Delta F\| &\leq C_3 \rho \end{aligned} \tag{III.9.3}$$

for constants C_1, C_2, C_3 only depending on the dimensions of the problem, and with

$$\rho = 2^{-t}$$

the relative computer precision [27].

We summarize (III.9.3) by writing

$$\Delta x = O(\rho) \qquad \Delta d = O(\rho) \qquad \Delta F = O(\rho)$$

We will always, for a given F, define the data vector so that (III.9.2) holds and so that the condition-number (see definition III.9.1) is minimized. Let $f\ell(d)$ denote the t digit binary representation of the vector d in floating-point arithmetic

$$\|f\ell(d) - d\| \leq C\rho\|d\| \quad \text{i.e. } f\ell(d) - d = O(\rho)$$

Since d is represented by $f\ell(d)$, we solve in fact $F(x;f\ell(d)) = 0$ instead of $F(x) = 0$, independent of the method used to solve (III.9.1). Let F'_x and F'_d denote the partial Fréchet-derivatives of F, respectively with respect to x and d.
Now $F(x;f\ell(d)) = 0$ has a root $\tilde{x}^*$ in the neighbourhood of x^* and $\tilde{x}^* - x^* = O(\rho)$ if t is sufficiently large:

$$\begin{aligned}\tilde{x}^* - x^* &= -F'_x(x^*;\ d)^{-1}\ F'_d(x^*;\ d)(f\ell(d) - d)\\ &\quad + \text{higher order terms in } \tilde{x}^* - x^* \text{ and } f\ell(d) - d\\ &= -F'_x(x^*;\ d)^{-1}\ F'_d(x^*;\ d)(f\ell(d) - d) + O(\rho^2),\end{aligned}$$

For $x^* \neq 0$: $\|\tilde{x}^* - x^*\|/\|x^*\| \leq \|F'_x(x^*;\ d)^{-1}F'_d(x^*;d)\|\ C\rho\ \|d\|/\|x^*\| + O(\rho^2)$

Definition III.9.1.:

$\mathrm{Cond}(F;\ d) = \|F'_x(x^*;\ d)^{-1}\ F'_d(x^*;\ d)\|.\|d\|/\|x^*\|$ is called the <u>condition number of F</u> with respect to the data vector d.

We call a problem ill-conditioned if $\mathrm{cond}(F;d) \gg 1$.
Let us now suppose that $F(x;d) = 0$ is solved by an iterative procedure $\Phi(x_i,F)$, where Φ can use several $F_i^{(j)}$, the jth Fréchet-derivative of F at x_i (if j=1 or 2, a single or double prime is used instead of the superscript j). If $\{x_i\}$ is the sequence of successive approximations of x^*, we can at best expect x_i to be the representation of a computed value for $\tilde{x}^*$,

$$\|x_i - \tilde{x}^*\| \leq K\rho\|x^*\|$$

So

$$\begin{aligned}\|x_i - x^*\| &\leq \|x_i - \tilde{x}^*\| + \|\tilde{x}^* - x^*\| \leq K\rho\|\tilde{x}^*\| + C\rho\ \mathrm{cond}(F;\ d).\|x^*\| + O(\rho^2)\\ &\leq K\rho(\|\tilde{x}^* - x^*\| + \|x^*\|) + C\rho\ \mathrm{cond}(F;\ d).\|x^*\| + O(\rho^2)\\ &\leq [K\rho + C\rho\ \mathrm{cond}(F;\ d)].\|x^*\| + O(\rho^2).\end{aligned}$$

Definition III.9.2.:

An iteration Φ is called <u>numerically stable</u> if

$$\lim_{i\to\infty} \|x_i - x^*\| \leq \rho.\|x^*\|.(C\ \mathrm{cond}(F;\ d) + K) + O(\rho^2),$$

with C and K nonnegative constants.

In practice we often want to find an approximation x_i such that $\|x_i - x^\star\| \le \varepsilon \cdot \|x^\star\|$. This is possible if the problem is sufficiently well-conditioned, i.e. $\rho\, \mathrm{cond}(F;\ d) = O(\varepsilon)$. In floating-point arithmetic we have

$$x_{i+1} = \Phi(x_i, F) + \xi_i \text{ where } \xi_i = f\ell(\Phi(x_i, F)) - \Phi(x_i, F)$$

Theorem III.9.1.:

A convergent iterative procedure $\Phi(x_i, F)$, i.e. $\lim\limits_{i\to\infty} \|\Phi(x_i, F) - x^\star\| = o$, is numerically stable if $\lim\limits_{i\to\infty} \|\xi_i\| \le \rho\|x^\star\| . (C.\mathrm{cond}(F;\ d) + K) + O(\rho^2)$

Proof:

We simply verify the definition:

$$\lim_{i\to\infty} \|x_i - x^\star\| \le \lim_{i\to\infty} [\|\Phi(x_{i-1}, F) - x^\star\| + \|\xi_{i-1}\|]$$

$$= \lim_{i\to\infty} \|\xi_{i-1}\| \le \rho\, \|x^\star\| (C\ \mathrm{cond}\ (F;\ d) + K) + O(\rho^2)$$

∎

9.2. *The Halley-iteration*

In [48] Woźniakowski proves numerical stability of the Newton-iteration for the solution of a system of nonlinear equations,

$$x_{i+1} = x_i + a_i$$

with

$$a_i = -F_i'^{-1} F_i$$

under a natural assumption on the computed evaluation of F.

Theorem III.9.2.:

If a) $f\ell(F(x_i;\ d)) = (I_x + \Delta F_i)\ F(x_i + \Delta x_i;\ d + \Delta d_i) = F(x_i;\ d) + \delta F_i$ with $\delta F_i = \Delta F_i\ F(x_i;\ d) + F_x'\ (x_i;\ d)\ \Delta x_i + F_d'\ (x_i; d)\ \Delta d_i + O(\rho^2)$

b) $f\ell(F'(x_i;\ d)) = F'(x_i;\ d) + \delta F_i'$ with $\delta F_i' = O(\rho)$

c) the computed correction $f\ell(a_i)$ is the exact solution of a perturbed linear system

$(F'(x_i;\ d) + \delta F_i' + E_i)\ f\ell(a_i) = -\ F(x_i;\ d) - \delta F_i$ with

$E_i = O(\rho)$

then the Newton-iteration is numerically stable

We will now prove numerical stability of the Halley-iteration for the solution of a system of nonlinear equations:

$$x_{i+1} = x_i + \frac{a_i^2}{a_i + \frac{1}{2} b_i} \tag{III.9.4}$$

with

$$b_i = F_i'^{-1} \; F_i'' \; a_i^2$$

under assumptions similar to the assumptions for the Newton-iteration. We will also assume that the divisions in (III.9.4) are such that

$$\left(\frac{1}{a_i + \frac{1}{2} F_i'^{-1} \; F_i'' \; a_i^2} \right)^j \; O(\|a_i\|^{j-k} \rho^{k+\ell}) = O(\rho^{\ell}) \tag{III.9.5}$$

Condition (III.9.5) takes care of the fact that the denominator of the correction-term in (III.9.4) does not become too small in comparison with $O(\|a_i\|^{j-k} \rho^k)$.
The assumption (III.9.5) is a natural generalization of the following relations:

$$\text{for } p = 1 : \lim_{i \to \infty} \frac{a_i}{a_i + \frac{1}{2} F_i'^{-1} \; F_i'' \; a_i^2} = 1$$

$$\Downarrow$$

$$\exists L \in \mathbb{N}, \; D \in \mathbb{R}_o^+ \ni \forall i \geq L : \left| \frac{a_i}{a_i + \frac{1}{2} F_i'^{-1} \; F_i'' \; a_i^2} \right| \leq 1 + D$$

(case $j = 1$, $k = 0$, $\ell = 0$)

and

in a convergent process (III.9.4) $\qquad \lim_{i \to \infty} \|x^* - x_i\| = o$

$$\lim_{i \to \infty} \frac{a_i^2}{a_i + \frac{1}{2} F_i'^{-1} \; F_i'' \; a_i^2} = 0 \qquad \qquad \lim_{i \to \infty} a_i = 0$$

$$\exists N \in \mathbb{N} \ni \forall i \geq N : \frac{a_i^2}{a_i + \frac{1}{2} F_i'^{-1} \; F_i'' \; a_i^2} = O(\rho) \qquad \exists M \in \mathbb{N} \ni \forall i \geq M : a_i = O(\rho)$$

$$\Downarrow$$

$$\exists M \in \mathbb{N} \ni \forall i \geq M : a_i^2 = O(\|a_i\| \rho)$$

$$\forall i \geq \max(N,M) : \frac{a_i^2}{a_i + \frac{1}{2} F_i'^{-1} \; F_i'' \; a_i^2} = \frac{1}{a_i + \frac{1}{2} F_i'^{-1} \; F_i'' \; a_i^2} \; O(\|a_i\| \rho) = O(\rho)$$

(case $j = 1$, $k = 0$, $\ell = 1$)

From now on we will sometimes write $F(x_i)$, $F'_x(x_i)$, $F'_d(x_i)$, $F'(x_i)$, $F''(x_i)$ instead of $F(x_i;d)$, $F'_x(x_i;d)$, $F'_d(x_i;d)$, $F'(x_i;d)$, $F''(x_i;d)$ in order to shorten the notations.

Theorem III.9.3.:

If a) $f\ell(F(x_i;d)) = (I_x+\Delta F_i)F(x_i+\Delta x_i;d+\Delta d_i) = F(x_i) + \delta F_i$ with

$\delta F_i = \Delta F_i F(x_i) + F'_x(x_i)\ \Delta x_i + F'_d(x_i)\ \Delta d_i + O(\rho^2)$

b) $f\ell(F'(x_i;d)) = F'(x_i) + \delta F'_i$ with $\delta F'_i = O(\rho)$

c) $f\ell(F''(x_i;d)) = F''(x_i) + \delta F''_i$ with $\delta F''_i = O(\rho)$

d) the computed correction $f\ell(a_i)$ is the exact solution of a perturbed linear system

$(F'(x_i) + \delta F'_i + E_{i,1})f\ell(a_i) = -F(x_i) - \delta F_i$, with $E_{i,1} = O(\rho)$

e) analogously,

$(F'(x_i) + \delta F'_i + E_{i,2})f\ell(b_i) = (F''(x_i) + \delta F''_i)\ f\ell(a_i)^2$ with $E_{i,2} = O(\rho)$

and (III.9.5) holds,

then the iteration (III.9.4) is numerically stable.

Proof:

Let $F'(x_i) + \delta F'_i + E_{i,1} = F'(x_i)(I_x+H_{i,1})$

where $H_{i,1} = F'(x_i)^{-1}\{\delta F'_i+E_{i,1}\} = O(\rho)$ because of b) and d).

So for small ρ

$$(I_x+H_{i,1})^{-1} = I_x-H_{i,1} + O(\rho^2)$$

Thus

$$f\ell(a_i) = (I_x-H_{i,1})F_i'^{-1}(-F_i-\delta F_i) + O(\rho^2) \qquad \text{(III.9.6)}$$

Analgously

$$f\ell(b_i) = (I_x-H_{i,2})F_i'^{-1}(F''_i+\delta F''_i)\ f\ell(a_i)^2 + O(\rho^2)$$

with $H_{i,2} = O(\rho)$.

Now

$$\begin{aligned}(F''_i+\delta F''_i)f\ell(a_i)^2 &= (F''_i+\delta F''_i)[(I_x-H_{i,1})F_i'^{-1}(-F_i-\delta F_i)]^2 + O(\rho^2)\\ &= (F''_i+\delta F''_i)a_i^2 + 2(F''_i+\delta F''_i)(F_i'^{-1}F_i,F_i'^{-1}\delta F_i-H_{i,1}F_i'^{-1}F_i) + O(\rho^2)\\ &= (F''_i+\delta F''_i)a_i^2 - 2F''_i(a_i,F_i'^{-1}\delta F_i-H_{i,1}F_i'^{-1}F_i) + O(\rho^2)\end{aligned}$$

Thus

$$f\ell(b_i) = F_i'^{-1}(F_i''+\delta F_i'')a_i^2 - 2F_i'^{-1}F_i''(a_i,F_i'^{-1}\delta F_i-H_{i,1}F_i'^{-1}F_i)$$
$$- H_{i,2}F_i'^{-1}F_i''\,a_i^2 + O(\rho^2)$$

A computed approximation x_{i+1} satisfies

$$x_{i+1} = (I_X+\delta I_{i,1})\left[x_i + (I_X+\delta I_{i,2})\frac{f\ell(a_i)^2}{f\ell(a_i)+\frac{1}{2}f\ell(b_i)}\right],$$

where $\delta I_{i,1}$ and $\delta I_{i,2}$ are diagonal matrices and $\delta I_{i,1} = O(\rho)$ and $\delta I_{i,2} = O(\rho)$. So

$$x_{i+1} = (I_X+\delta I_{i,1})\left[x_i + (I_X+\delta I_{i,2})\frac{a_i^2-2a_i\cdot(F_i'^{-1}\delta F_i+H_{i,1}a_i) + O(\rho^2)}{a_i+\frac{1}{2}b_i - \delta a_i + O(\rho^2)}\right],$$

where

$$\delta a_i = F_i'^{-1}\delta F_i + H_{i,1}a_i - \tfrac{1}{2}F_i'^{-1}\delta F_i''\,a_i^2$$
$$+\tfrac{1}{2}H_{i,2}F_i'^{-1}F_i''\,a_i^2 + F_i'^{-1}F_i''\,(a_i,\,F_i'^{-1}\,\delta F_i - H_{i,1}F_i'^{-1}F_i).$$

Using (III.9.6) we find

$$f\ell(a_i) - a_i + H_{i,1}\,a_i - H_{i,1}F_i'^{-1}\delta F_i + O(\rho^2) = -F_i'^{-1}\delta F_i,$$

and thus, for positive constants D_1 and D_2,

$$\|F_i'^{-1}\delta F_i\| \le D_2\rho\|a_i\|$$

since

$$\|f\ell(a_i)-a_i\| \le D_1\rho\|a_i\|$$

and

$$\|F_i'^{-1}\|\cdot\|F_i\| \le \|F_i'^{-1}\|\cdot\|F_i'\|\cdot\|a_i\|.$$

Thus

$$x_{i+1} = (I_X+\delta I_{i,1})\left[x_i + \frac{a_i^2 - 2a_i(F_i'^{-1}\delta F_i+H_{i,1}a_i) + \delta I_{i,2}a_i^2 + O(\rho^2\|a_i\|^2,\rho^2)}{a_i+\frac{1}{2}b_i - \delta a_i + O(\rho^2)}\right]$$

where $\delta I_{i,2}\, a_i^2$ is the linear operator $\delta I_{i,2}$ evaluated in a_i^2 (componentwise square of the vector a_i).

So

$$x_{i+1} = (I_x+\delta I_{i,1})\left[x_i + \frac{a_i^2 - 2a_i(F_i'^{-1}\delta F_i + H_{i,1}a_i) + \delta I_{i,2}a_i^2 + O(\rho^2\|a_i\|^2,\rho^2)}{a_i + \frac{1}{2}b_i}\right.$$

with

$$c_i = I + \frac{1}{a_i + \frac{1}{2}b_i}(\delta a_i + O(\rho^2)) + \left(\frac{1}{a_i + \frac{1}{2}b_i}\right)^2 O(\|a_i\|^{2-k}\rho^{k+2},\ k=0,1,2)$$

since $\delta a_i = O(\rho\|a_i\|)$, where I is the unit vector $(1,\dots,1)$ in $\mathbb{R}^p$.

Using (III.9.5) we conclude

$$\left(\frac{1}{a_i + \frac{1}{2}b_i}\right)^2 O(\|a_i\|^{2-k}\rho^{k+2},\ k=0,1,2) = O(\rho^2).$$

For $\xi_i = x_{i+1} - \Phi(x_i,F)$, we have

$$\xi_i = \delta I_{i,1}\, x_i + \frac{a_i^2}{a_i + \frac{1}{2}b_i}(c_i - I)$$

$$+ \frac{-2a_i(F_i'^{-1}\delta F_i + H_{i,1}a_i) + \delta I_{i,2}a_i^2 + O(\rho^2\|a_i\|^2)}{a_i + \frac{1}{2}b_i}.c_i$$

$$+ \delta I_{i,1}\frac{a_i^2}{a_i + \frac{1}{2}b_i}.c_i + O(\rho^2).$$

So

$$\xi_i = \delta I_{i,1}\, x_i + \left(\frac{1}{a_i + \frac{1}{2}b_i}\right)^2 O(\rho\|a_i\|^3,\rho^2\|a_i\|^2)$$

$$+ \frac{1}{a_i + \frac{1}{2}b_i}(-2a_iF_i'^{-1}\delta F_i + O(\rho\|a_i\|^2,\rho^2\|a_i\|^2)).(1 + O(\rho))$$

$$+ O(\rho^2).$$

Thus

$$\|\xi_i\| \le k_1\rho\|x_i\| + k_2\rho\|a_i\| + \left\|\frac{-2a_i}{a_i + \frac{1}{2}b_i}F_i'^{-1}\delta F_i\right\| + O(\rho^2),$$

and since

$$\frac{-2a_i}{a_i+\frac{1}{2}b_i} F_i'^{-1}\delta F_i = \frac{-2a_i}{a_i+\frac{1}{2}b_i} F_i'^{-1}(\Delta F_i F(x_i) + F_x'(x_i)\Delta x_i + F_d'(x_i)\Delta d_i) + O(\rho^2))$$

$$= \frac{1}{a_i+\frac{1}{2}b_i} O(\rho\|a_i\|)F(x_i) - \frac{2a_i}{a_i+\frac{1}{2}b_i}\Delta x_i$$

$$- \frac{2a_i}{a_i+\frac{1}{2}b_i} F_i'^{-1}F_d'(x_i)\Delta d_i + \frac{1}{a_i+\frac{1}{2}b_i} O(\rho^2\|a_i\|),$$

we find that

$$\lim_{i\to\infty} \|\xi_i\| \le \rho\|x^\star\|(K + C\ \mathrm{cond}(F;d)) + O(\rho^2)$$

for $\lim_{i\to\infty} a_i = 0 = \lim_{i\to\infty} F(x_i)$ in a convergent process and $a_i\Delta x_i = O(\rho\|a_i\|)$ and $a_i F_i'^{-1}F_d'(x_i)\Delta d_i = O(\rho\|a_i\|)$.

■

9.3. *Example*

Consider the following operator

$$F : \mathbb{R}^2 \to \mathbb{R}^2 : \binom{x}{y} \to \begin{pmatrix} e^{-x+y} - d_1 \\ e^{-x-y} - d_2 \end{pmatrix} \text{ with } d_1 > 0 \text{ and } d_2 > 0.$$

The operator F has a simple root $x^\star = (-\frac{1}{2}\ln(d_1 d_2), \frac{1}{2}\ln(d_1/d_2))$.

Clearly $d = (d_1, d_2) \in \mathbb{R}^2$ is the data vector

Now

$$fl(F(x,y;d)) = \begin{pmatrix} [(1+\varepsilon_1)e^{(-x-\Delta'x+y+\Delta'y)(1+\theta_1)} - (d_1+\Delta_1'd)]\ (1+\kappa_1) \\ [(1+\varepsilon_2)e^{(-x-\Delta'x-y-\Delta'y)(1+\theta_2)} - (d_2+\Delta_2'd)]\ (1+\kappa_2) \end{pmatrix}$$

where $fl(x) = x + \Delta'x$, $fl(y) = y + \Delta'y$, $fl(d_1) = d_1 + \Delta_1'd$, $fl(d_2) = d_2 + \Delta_2'd$, θ_1 is caused by $-fl(x) + fl(y)$, θ_2 is caused by $-fl(x) - fl(y)$, ε_i are caused by the exponential evaluations $(i=1,2)$, κ_i are caused by the subtraction of $fl(d_i)$ $(i=1,2)$. One can rewrite $fl(F(x,y;d)) = (I_x+\Delta F)F(x+\Delta x, y+\Delta y; d+\Delta d)$ with

$$\Delta x = x\theta_1 + \Delta' x(1+\theta_1),\ \Delta y = y\theta_1 + \Delta' y(1+\theta_1),\ \Delta d = (\Delta_1 d,\ \Delta_2 d),$$

$$\Delta_1 d = \frac{\Delta_1' d - \varepsilon_1 d_1}{1 + \varepsilon_1},$$

$$\Delta_2 d = \frac{\Delta_2' d - \varepsilon_2 d_2}{1 + \varepsilon_2} + \frac{d_2 + \Delta_2' d}{1 + \varepsilon_2}\left(e^{(x+\Delta' x+y+\Delta' y)(\theta_2-\theta_1)} - 1\right),$$

$$\Delta F = \begin{pmatrix} (1+\varepsilon_1)(1+\kappa_1)-1 & 0 \\ 0 & (1+\varepsilon_2)(1+\kappa_2)e^{(x+\Delta' x+y+\Delta' y)(\theta_1-\theta_2)}-1 \end{pmatrix}$$

The inverse of the Jacobian matrix in the root x^* is

$$\frac{1}{2(d_1 \cdot d_2)}\begin{pmatrix} -d_2 & -d_1 \\ d_2 & -d_1 \end{pmatrix} \text{ and } F_d' = \begin{pmatrix} -1 & 0 \\ 0 & -1 \end{pmatrix}$$

The condition-number of F with respect to the data vector d is

$$\|F_x'(x^*;d)^{-1}\| \cdot \frac{\|(d_1,d_2)\|}{\|x^*\|}$$

Using the Schur-norm $\|A\| = \sqrt{\Sigma_{i,j}\, a_{ij}^2}$ of a matrix $A = (a_{ij})$ and the L_2-norm $\|a\| = \sqrt{\Sigma_i\, a_i^2}$ of a vector $a = (a_i)$, the condition-number is

$$\frac{d_1^2 + d_2^2}{\sqrt{2}\, d_1 d_2 \|x^*\|}$$

Putting $d_1 = d = d_2$, the root $x^* = (-\ln d, 0)$ and the condition-number is $\sqrt{2}/|\ln d|$.
The problem is extremely well-conditioned if $\mathrm{cond}(F;d) \leq 1$, i.e.

$$d \in\ ]-\infty,\ e^{-\sqrt{2}}]\ \cup\ [e^{\sqrt{2}},\ +\infty[$$

The problem is very ill-conditioned if $d = e^{\varepsilon}$ with ε very small.
We will now check some of the conditions of theorem III.9.3. We already know
$fl(F(x,y;d)) = (I_x+\Delta F)F(x+\Delta x, y+\Delta y; d+\Delta d)$. Now

$$fl(F'(x,y;d)) = fl\begin{pmatrix} -e^{-x+y} & e^{-x+y} \\ -e^{-x-y} & -e^{-x-y} \end{pmatrix}$$

where

$$fl(e^{-x+y}) = (1+\varepsilon_1)\, e^{(-x-\Delta'x+y+\Delta'y)(1+\theta_1)} = (1+\varepsilon_1)\, e^{-x+y}\, e^{-\Delta x+\Delta y}$$
$$= e^{-x+y}\, [\, 1+\varepsilon_1+(1+\varepsilon_1)(e^{-\Delta x+\Delta y} -1)] ,$$

and

$$fl(e^{-x-y}) = (1+\varepsilon_2)\, e^{(-x-\Delta'x-y-\Delta'y)(1+\theta_2)}$$
$$= (1+\varepsilon_2)\, e^{-x-y} e^{-\Delta x-\Delta y}\, e^{(x+\Delta'x+y+\Delta'y)(\theta_1-\theta_2)}$$
$$= e^{-x-y}\, [\, 1+\varepsilon_2+(1+\varepsilon_2)(e^{-\Delta x-\Delta y}\, e^{(x+\Delta'x+y+\Delta'y)(\theta_1-\theta_2)} -1)] .$$

So $fl(F'(x,y;d)) = F'(x,y;d) + \delta F'(x,y;d)$ with

$\delta F'(x,y;d)$

$$= \begin{pmatrix} \varepsilon_1+(1+\varepsilon_1)(e^{-\Delta x+\Delta y} - 1) & 0 \\ 0 & \varepsilon_2+(1+\varepsilon_2)(e^{-\Delta x-\Delta y}\, e^{(x+\Delta'x+y+\Delta'y)(\theta_1-\theta_2)} - 1) \end{pmatrix} . F'(x,y;d)$$

$= O(\rho)$

We can write down an analogous formula for $F''(x,y;d)$.

The two linear systems of equations are well-conditioned since the condition-number of the linear systems in $x^* = \lim_{i\to\infty} x_i$ is

$$\|F'_x\,(x^*;d)^{-1}\| . \|F'_x(x^*;d)\| = 2$$

One can prove that the use of Gaussian elimination with row pivoting for this example satisfies the conditions d) and e) of theorem III.9.3. So we can expect to get a reasonable approximation of the solution of $F(x,y;d) = 0$ using the numerically stable iterative method (III.9.4); the numerical results illustrate this. Let us at the same time follow the loss of significant digits in the root x^* as the problem becomes worse-conditioned. The calculations are performed in double precision (t=56). We solve the nonlinear system $F(x,y;d) = 0$ for $d = \exp(10^{-k})$, $k=0,\ldots,16$. The root $x^* = (-10^{-k},0)$. In table III.9.1 we give for each d the 6th iterationstep (x_6,y_6) in the procedure (III.9.4) starting from $(x_0,y_0) = (2,2)$, the number ℓ of significant digits in x_6, and the condition-number cond $(F\ ;\ \exp(10^{-k})\ )$.

It is also important to know that the iterative procedure stops at the 6th iterationstep, except for k=7, 13 and 14 where respectively ℓ=11, 5 and 3 in the last iterationstep (x_7,y_7). We have used the stop-criterion

$$\max(|x_{i+1} - x_i|,\ |y_{i+1} - y_i|) \le 10^{-15} \max(|x_{i+1}|,\ |y_{i+1}|).$$

We remark that the algorithm even behaves considerably well for a condition-number of the order of 10^3 or 10^4.

k	x_6	y_6	ℓ	$\mathrm{cond}(F;e^{10^{-k}})$
0	-0.1000000000000000(01)	0.3597855161523896(-18)	16	$\sqrt{2}$
1	-0.1000000000000000(00)	-0.2376055789464463(-17)	16	$10\sqrt{2}$
2	-0.1000000000000001(-01)	-0.6397150159689099(-17)	15	$10^2\sqrt{2}$
3	-0.0999999999999997(-02)	0.5077502606368951(-17)	15	$10^3\sqrt{2}$
4	-0.0999999999999844(-03)	0.3913464269882279(-17)	13	$10^4\sqrt{2}$
5	-0.0999999999997470(-04)	-0.3905797959965137(-17)	12	$10^5\sqrt{2}$
6	-0.0999999999986935(-05)	0.5633677343553680(-17)	11	$10^6\sqrt{2}$
7	-0.1000000000174599(-06)	-0.1058449777227516(-16)	10	$10^7\sqrt{2}$
8	-0.1000000000015281(-07)	0.4124494865312562(-17)	11	$10^8\sqrt{2}$
9	-0.1000000007452433(-08)	-0.2449359520991520(-17)	9	$10^9\sqrt{2}$
10	-0.0999999914314586(-09)	0.4265833288825851(-17)	8	$10^{10}\sqrt{2}$
11	-0.1000000261210709(-10)	-0.6446772724219823(-17)	7	$10^{11}\sqrt{2}$
12	-0.0999980430668081(-11)	0.3302303528672576(-17)	5	$10^{12}\sqrt{2}$
13	-0.0999761308551817(-12)	0.1322187990417560(-16)	4	$10^{13}\sqrt{2}$
14	-0.1000372750236664(-13)	-0.1182870095748150(-16)	4	$10^{14}\sqrt{2}$
15	-0.0963108239652912(-14)	0.1398012990192197(-17)	2	$10^{15}\sqrt{2}$
16	-0.0868560967896870(-15)	0.3349523961106902(-17)	1	$10^{16}\sqrt{2}$

Table III.9.1.

REFERENCES

[1] Abramowitz M. - Handbook of mathematical functions with formulas, graphs and mathematical tables. New York, 1968.

[2] Baker G. A. - Essentials of Padé approximants Academic Press, New York, 1975.

[3] Basu S. & Bose N.K. - Two-dimensional matrix Padé-approximants: existence, nonuniqueness and recursive computation. IEEE Transactions Autom. Control 25(3), 1980, pp. 509-514.

[4] Bessis J.D. & Talman J.D. - Variational approach to the theory of operator Padé-approximants. Rocky Mountain Journ. Math. 4, 1974, pp. 151-158.

[5] Birkhoff G. & Mc Lane S. - Algebra Collier - Mc Millan Ltd, London, 1967.

[6] Bögel K. & Tasche M. - Analysis in normierten Raümen Akademie Verlag, Berlin, 1974.

[7] Bose N.K. - Two-dimensional approximants via one-dimensional Padé-technique Signal Processing: theories and applcs, 1980, pp. 409-411.

[8] Brezinski C. - Accélération de la convergence en analyse numérique. L.N.M. 584, Springer Verlag, Berlin, 1977.

[9] Brezinski C. - Algoritmes d'accélération de la convergence Editions Technip, Paris, 1978.

[10] Chandrasekhar S. - Radiative Transfer. New York, 1960.

[11] Chisholm J.S.R. - Multivariate approximants with branch points: diagonal approximants. Proc. Royal Soc. Lond. A 358, 1977, pp. 351-366

[12] Chisholm J.S.R. - Multivariate approximants with branch points: off-diagonal approximants Proc. Royal Soc. Lond. A 362, 1978, pp. 43-56

[13] Chisholm J.S.R. - N-variable rational approximants. in [44], pp. 23-42

[14] Chisholm J.S.R. - Rational approximants defined from double power series. Math. Comp. 27, 1973, pp. 841-848.

[15] Döring B. - Ein Satz über eine von Grebenjuk betrachtete Klasse von Iterationsverfahren.
Iterationsverfahren, Numerische Mathematik und Approximationstheorie, Birkhäuser Verlag, Basel, 1970, pp. 195-226

[16] Ehrmann H. - Konstruktion und Durchführung von Iterationsverfahren höherer Ordnung.
Arch. Rat. Mech. Anal. 4, 1959-1960, pp. 65-88

[17] Fair W. & Luke Y.L. - Padé-approximants to the operator exponential
Num. Math. 14, 1970, pp. 379-382.

[18] Frame J.S. - The solution of equations by continued fractions.
Amer. Math. Monthly 60, 1953, pp. 293-305

[19] Friedlander B. & Kailath T. & Ljung L. & Morf M. -
New inversion formulas for matrices classified in terms of their distance form Toeplitz-matrices.
Lin. Algebra Applcs 27, 1979, pp. 31-60.

[20] Frobenius G. - Uber relationen zwischen den Näherungsbrüchen von Potenzreihen.
Journ. Reine Angewandte Math. 90, 1881, pp. 1-17

[21] Gammel J.L. - Review of two recent generalizations of the Padé-Approximant
in [24], pp. 3-9

[22] Genz A. - The approximate calculation of multidimensional integrals using extrapolation methods.
Ph. D. in Applied Mathematics, University of Kent, 1975.

[23] Godement R. - Algebra
Kershaw Publ. Co. Ltd, London, 1969.

[24] Graves-Morris P.R. - Padé-approximants and their applications
Academic Press, London and New York, 1973.

[25] Graves-Morris P. & Hughes Jones R. & Makinson G. -
The calculation of some rational approximants in two variables.
Journ. Inst. Math. Applcs 13, 1974, pp. 311-320

[26] Greenspan D. - Introductory numerical analysis of elliptic boundary value problems.
Harper and Row, New York, 1966.

[27] Gregory R. & Young D. - A survey of Numerical Mathematics I.
Addison-Wesley, Reading Massachusetts, 1972.

[28] Hillion P. - Remarks on rational approximations of multiple power series.
Journ. Inst. Math. Applcs 19, 1977, pp. 281-293

[29] Hughes Jones R. - General Rational approximants in N variables
Journ. Approx. Theory 16, 1976, pp. 201-233.

[30] Hughes Jones R. & Makinson G.J.- The generation of Chisholm rational polynomial approximants to power series in two variables.
Journ. Inst. Math. Applcs 13, 1974, pp.299-310.

[31] Kailath T. & Kung S.Y. & Morf M. - Displacement ranks of matrices and linear equations.
Journ. Math. Anal. Applcs 68, 1979, pp. 395-407.

[32] Karlsson J. & Wallin H. - Rational approximation by an interpolation procedure in several variables.
in [44], pp. 83-100.

[33] Larsen R. - Banach algebras, an introduction.
Marcel Dekker, New York, 1973.

[34] Levin D. - General order Padé-type rational approximants defined from double power series.
Journ. Inst. Math. Applcs 18, 1976, pp. 1-8.

[35] Levin D. - On accelerating the convergence of infinite double series and integrals.
Math. Comp. 35(152), 1980, pp. 1331-1345.

[36] Lutterodt C.H. - A two-dimensional analogue of Padé-approximant theory.
Journ. Phys. A Math. 7(9), 1974, pp. 1027-1037

[37] Lutterodt C.H. - Rational approximants to holomorphic functions in n dimensions.
Journ. Math. Anal. Applcs 53, 1976, pp. 89-98.

[38] Ortega J.M. - Numerical Analysis, a second course.
Academic Press, New York, 1972.

[39] Ortega J.M. & Rheinboldt W.C. - Iterative solution of nonlinear equations in several variables
Academic Press, New York and London, 1970.

[40] Pohozaev S.J. - The Dirichlet problem for the equation $\Delta u = u^2$
Soviet Math. Dokl. 1, 1960, pp. 1143-1146

[41] Rall L.B. - Computational Solution of nonlinear operator equations
Krieger Huntington, New York, 1979.

[42] Rall L.B. - Quadratic equations in Banach Spaces
Rend. Circ. Mat. Palermo 10, 1961, pp. 314-332.

[43] Rudin W. - Functional Analysis
Mc Graw Hill, New York, 1973

[44] Saff E.B. & Varga R.S. - Padé and rational approximation: theory and applications
Academic Press, London, 1977.

[45] Shafer R.E. - On quadratic approximation
Siam Journ. Num. Anal. 11(2), 1974, pp. 447-460.

[46] Starkand Y. - Explicit formulas for matrix-valued Padé-approximants.
Journ. Comp. Appl. Math. 5(1), 1979, pp. 63-66.

[47] Stibbs D.W. & Weir R.E. - On the H-functions for isotropic scattering.
Monthly Not. Royal Astron. Soc. 119, 1959, pp. 512-525

[48] Woźniakowski H. - Numerical stability for solving nonlinear equations.
Num. Math. 27, 1977, pp. 373-390.

[49] Wuytack L. - Applications of Padé-approximation in numerical analysis.
Approximation Theory (Schabach R. & Scherer K. eds)
Springer, Berlin, 1976, pp. 453-461.

SUBJECT INDEX

Vol. 926: Geometric Techniques in Gauge Theories. Proceedings, 1981. Edited by R. Martini and E.M.de Jager. IX, 219 pages. 1982.

Vol. 927: Y. Z. Flicker, The Trace Formula and Base Change for GL (3). XII, 204 pages. 1982.

Vol. 928: Probability Measures on Groups. Proceedings 1981. Edited by H. Heyer. X, 477 pages. 1982.

Vol. 929: Ecole d'Eté de Probabilités de Saint-Flour X – 1980. Proceedings, 1980. Edited by P.L. Hennequin. X, 313 pages. 1982.

Vol. 930: P. Berthelot, L. Breen, et W. Messing, Théorie de Dieudonné Cristalline II. XI, 261 pages. 1982.

Vol. 931: D.M. Arnold, Finite Rank Torsion Free Abelian Groups and Rings. VII, 191 pages. 1982.

Vol. 932: Analytic Theory of Continued Fractions. Proceedings, 1981. Edited by W.B. Jones, W.J. Thron, and H. Waadeland. VI, 240 pages. 1982.

Vol. 933: Lie Algebras and Related Topics. Proceedings, 1981. Edited by D. Winter. VI, 236 pages. 1982.

Vol. 934: M. Sakai, Quadrature Domains. IV, 133 pages. 1982.

Vol. 935: R. Sot, Simple Morphisms in Algebraic Geometry. IV, 146 pages. 1982.

Vol. 936: S.M. Khaleelulla, Counterexamples in Topological Vector Spaces. XXI, 179 pages. 1982.

Vol. 937: E. Combet, Intégrales Exponentielles. VIII, 114 pages. 1982.

Vol. 938: Number Theory. Proceedings, 1981. Edited by K. Alladi. IX, 177 pages. 1982.

Vol. 939: Martingale Theory in Harmonic Analysis and Banach Spaces. Proceedings, 1981. Edited by J.-A. Chao and W.A. Woyczyński. VIII, 225 pages. 1982.

Vol. 940: S. Shelah, Proper Forcing. XXIX, 496 pages. 1982.

Vol. 941: A. Legrand, Homotopie des Espaces de Sections. VII, 132 pages. 1982.

Vol. 942: Theory and Applications of Singular Perturbations. Proceedings, 1981. Edited by W. Eckhaus and E.M. de Jager. V, 363 pages. 1982.

Vol. 943: V. Ancona, G. Tomassini, Modifications Analytiques. IV, 120 pages. 1982.

Vol. 944: Representations of Algebras. Workshop Proceedings, 1980. Edited by M. Auslander and E. Lluis. V, 258 pages. 1982.

Vol. 945: Measure Theory. Oberwolfach 1981, Proceedings. Edited by D. Kölzow and D. Maharam-Stone. XV, 431 pages. 1982.

Vol. 946: N. Spaltenstein, Classes Unipotentes et Sous-groupes de Borel. IX, 259 pages. 1982.

Vol. 947: Algebraic Threefolds. Proceedings, 1981. Edited by A. Conte. VII, 315 pages. 1982.

Vol. 948: Functional Analysis. Proceedings, 1981. Edited by D. Butković, H. Kraljević, and S. Kurepa. X, 239 pages. 1982.

Vol. 949: Harmonic Maps. Proceedings, 1980. Edited by R.J. Knill, M. Kalka and H.C.J. Sealey. V, 158 pages. 1982.

Vol. 950: Complex Analysis. Proceedings, 1980. Edited by J. Eells. IV, 428 pages. 1982.

Vol. 951: Advances in Non-Commutative Ring Theory. Proceedings, 1981. Edited by P.J. Fleury. V, 142 pages. 1982.

Vol. 952: Combinatorial Mathematics IX. Proceedings, 1981. Edited by E. Billington, S. Oates-Williams, and A.P. Street. XI, 443 pages. 1982.

Vol. 953: Iterative Solution of Nonlinear Systems of Equations. Proceedings, 1982. Edited by R. Ansorge, Th. Meis, and W. Törnig. VII, 202 pages. 1982.

Vol. 954: S.G. Pandit, S.G. Deo, Differential Systems Involving Impulses. VII, 102 pages. 1982.

Vol. 955: G. Gierz, Bundles of Topological Vector Spaces and Their Duality. IV, 296 pages. 1982.

Vol. 956: Group Actions and Vector Fields. Proceedings, 1981. Edited by J.B. Carrell. V, 144 pages. 1982.

Vol. 957: Differential Equations. Proceedings, 1981. Edited by D.G. de Figueiredo. VIII, 301 pages. 1982.

Vol. 958: F.R. Beyl, J. Tappe, Group Extensions, Representations, and the Schur Multiplicator. IV, 278 pages. 1982.

Vol. 959: Géométrie Algébrique Réelle et Formes Quadratiques, Proceedings, 1981. Edité par J.-L. Colliot-Thélène, M. Coste, L. Mahé, et M.-F. Roy. X, 458 pages. 1982.

Vol. 960: Multigrid Methods. Proceedings, 1981. Edited by W. Hackbusch and U. Trottenberg. VII, 652 pages. 1982.

Vol. 961: Algebraic Geometry. Proceedings, 1981. Edited by J.M. Aroca, R. Buchweitz, M. Giusti, and M. Merle. X, 500 pages. 1982.

Vol. 962: Category Theory. Proceedings, 1981. Edited by K.H. Kamps, D. Pumplün, and W. Tholen, XV, 322 pages. 1982.

Vol. 963: R. Nottrot, Optimal Processes on Manifolds. VI, 124 pages. 1982.

Vol. 964: Ordinary and Partial Differential Equations. Proceedings, 1982. Edited by W.N. Everitt and B.D. Sleeman. XVIII, 726 pages. 1982.

Vol. 965: Topics in Numerical Analysis. Proceedings, 1981. Edited by P.R. Turner. IX, 202 pages. 1982.

Vol. 966: Algebraic K-Theory. Proceedings, 1980, Part I. Edited by R.K. Dennis. VIII, 407 pages. 1982.

Vol. 967: Algebraic K-Theory. Proceedings, 1980. Part II. VIII, 409 pages. 1982.

Vol. 968: Numerical Integration of Differential Equations and Large Linear Systems. Proceedings, 1980. Edited by J. Hinze. VI, 412 pages. 1982.

Vol. 969: Combinatorial Theory. Proceedings, 1982. Edited by D. Jungnickel and K. Vedder. V, 326 pages. 1982.

Vol. 970: Twistor Geometry and Non-Linear Systems. Proceedings, 1980. Edited by H.-D. Doebner and T.D. Palev. V, 216 pages. 1982.

Vol. 971: Kleinian Groups and Related Topics. Proceedings, 1981. Edited by D.M. Gallo and R.M. Porter. V, 117 pages. 1983.

Vol. 972: Nonlinear Filtering and Stochastic Control. Proceedings, 1981. Edited by S.K. Mitter and A. Moro. VIII, 297 pages. 1983.

Vol. 973: Matrix Pencils. Proceedings, 1982. Edited by B. Kågström and A. Ruhe. XI, 293 pages. 1983.

Vol. 974: A. Draux, Polynômes Orthogonaux Formels – Applications. VI, 625 pages. 1983.

Vol. 975: Radical Banach Algebras and Automatic Continuity. Proceedings, 1981. Edited by J.M. Bachar, W.G. Bade, P.C. Curtis Jr., H.G. Dales and M.P. Thomas. VIII, 470 pages. 1983.

Vol. 976: X. Fernique, P.W. Millar, D.W. Stroock, M. Weber, Ecole d'Eté de Probabilités de Saint-Flour XI – 1981. Edited by P.L. Hennequin. XI, 465 pages. 1983.

Vol. 977: T. Parthasarathy, On Global Univalence Theorems. VIII, 106 pages. 1983.

Vol. 978: J. Ławrynowicz, J. Krzyż, Quasiconformal Mappings in the Plane. VI, 177 pages. 1983.

Vol. 979: Mathematical Theories of Optimization. Proceedings, 1981. Edited by J.P. Cecconi and T. Zolezzi. V, 268 pages. 1983.

Vol. 980: L. Breen. Fonctions thêta et théorème du cube. XIII, 115 pages. 1983.

Vol. 981: Value Distribution Theory. Proceedings, 1981. Edited by I. Laine and S. Rickman. VIII, 245 pages. 1983.

Vol. 982: Stability Problems for Stochastic Models. Proceedings, 1982. Edited by V. V. Kalashnikov and V. M. Zolotarev. XVII, 295 pages. 1983.

Vol. 983: Nonstandard Analysis-Recent Developments. Edited by A. E. Hurd. V, 213 pages. 1983.

Vol. 984: A. Bove, J. E. Lewis, C. Parenti, Propagation of Singularities for Fuchsian Operators. IV, 161 pages. 1983.

Vol. 985: Asymptotic Analysis II. Edited by F. Verhulst. VI, 497 pages. 1983.

Vol. 986: Séminaire de Probabilités XVII 1981/82. Proceedings. Edited by J. Azéma and M. Yor. V, 512 pages. 1983.

Vol. 987: C. J. Bushnell, A. Fröhlich, Gauss Sums and p-adic Division Algebras. XI, 187 pages. 1983.

Vol. 988: J. Schwermer, Kohomologie arithmetisch definierter Gruppen und Eisensteinreihen. III, 170 pages. 1983.

Vol. 989: A. B. Mingarelli, Volterra-Stieltjes Integral Equations and Generalized Ordinary Differential Expressions. XIV, 318 pages. 1983.

Vol. 990: Probability in Banach Spaces IV. Proceedings, 1982. Edited by A. Beck and K. Jacobs. V, 234 pages. 1983.

Vol. 991: Banach Space Theory and its Applications. Proceedings, 1981. Edited by A. Pietsch, N. Popa and I. Singer. X, 302 pages. 1983.

Vol. 992: Harmonic Analysis, Proceedings, 1982. Edited by G. Mauceri, F. Ricci and G. Weiss. X, 449 pages. 1983.

Vol. 993: R. D. Bourgin, Geometric Aspects of Convex Sets with the Radon-Nikodým Property. XII, 474 pages. 1983.

Vol. 994: J.-L. Journé, Calderón-Zygmund Operators, Pseudo-Differential Operators and the Cauchy Integral of Calderón. VI, 129 pages. 1983.

Vol. 995: Banach Spaces, Harmonic Analysis, and Probability Theory. Proceedings, 1980–1981. Edited by R.C. Blei and S.J. Sidney. V, 173 pages. 1983.

Vol. 996: Invariant Theory. Proceedings, 1982. Edited by F. Gherardelli. V, 159 pages. 1983.

Vol. 997: Algebraic Geometry – Open Problems. Edited by C. Ciliberto, F. Ghione and F. Orecchia. VIII, 411 pages. 1983.

Vol. 998: Recent Developments in the Algebraic, Analytical, and Topological Theory of Semigroups. Proceedings, 1981. Edited by K.H. Hofmann, H. Jürgensen and H. J. Weinert. VI, 486 pages. 1983.

Vol. 999: C. Preston, Iterates of Maps on an Interval. VII, 205 pages. 1983.

Vol. 1000: H. Hopf, Differential Geometry in the Large, VII, 184 pages. 1983.

Vol. 1001: D.A. Hejhal, The Selberg Trace Formula for PSL(2, $\mathbb{R}$). Volume 2. VIII, 806 pages. 1983.

Vol. 1002: A. Edrei, E.B. Saff, R.S. Varga, Zeros of Sections of Power Series. VIII, 115 pages. 1983.

Vol. 1003: J. Schmets, Spaces of Vector-Valued Continuous Functions. VI, 117 pages. 1983.

Vol. 1004: Universal Algebra and Lattice Theory. Proceedings, 1982. Edited by R. S. Freese and O. C. Garcia. VI, 308 pages. 1983.

Vol. 1005: Numerical Methods. Proceedings, 1982. Edited by V. Pereyra and A. Reinoza. V, 296 pages. 1983.

Vol. 1006: Abelian Group Theory. Proceedings, 1982/83. Edited by R. Göbel, L. Lady and A. Mader. XVI, 771 pages. 1983.

Vol. 1007: Geometric Dynamics. Proceedings, 1981. Edited by J. Palis Jr. IX, 827 pages. 1983.

Vol. 1008: Algebraic Geometry. Proceedings, 1981. Edited by J. Dolgachev. V, 138 pages. 1983.

Vol. 1009: T. A. Chapman, Controlled Simple Homotopy Theory and Applications. III, 94 pages. 1983.

Vol. 1010: J.-E. Dies, Chaînes de Markov sur les permutations. IX, 226 pages. 1983.

Vol. 1011: J.M. Sigal. Scattering Theory for Many-Body Quantum Mechanical Systems. IV, 132 pages. 1983.

Vol. 1012: S. Kantorovitz, Spectral Theory of Banach Space Operators. V, 179 pages. 1983.

Vol. 1013: Complex Analysis – Fifth Romanian-Finnish Seminar. Part 1. Proceedings, 1981. Edited by C. Andreian Cazacu, N. Boboc, M. Jurchescu and I. Suciu. XX, 393 pages. 1983.

Vol. 1014: Complex Analysis – Fifth Romanian-Finnish Seminar. Part 2. Proceedings, 1981. Edited by C. Andreian Cazacu, N. Boboc, M. Jurchescu and I. Suciu. XX, 334 pages. 1983.

Vol. 1015: Equations différentielles et systèmes de Pfaff dans le champ complexe – II. Seminar. Edited by R. Gérard et J.P. Ramis. V, 411 pages. 1983.

Vol. 1016: Algebraic Geometry. Proceedings, 1982. Edited by M. Raynaud and T. Shioda. VIII, 528 pages. 1983.

Vol. 1017: Equadiff 82. Proceedings, 1982. Edited by H. W. Knobloch and K. Schmitt. XXIII, 666 pages. 1983.

Vol. 1018: Graph Theory, Łagów 1981. Proceedings, 1981. Edited by M. Borowiecki, J. W. Kennedy and M. M. Sysło. X, 289 pages. 1983.

Vol. 1019: Cabal Seminar 79–81. Proceedings, 1979–81. Edited by A. S. Kechris, D. A. Martin and Y. N. Moschovakis. V, 284 pages. 1983.

Vol. 1020: Non Commutative Harmonic Analysis and Lie Groups. Proceedings, 1982. Edited by J. Carmona and M. Vergne. V, 187 pages. 1983.

Vol. 1021: Probability Theory and Mathematical Statistics. Proceedings, 1982. Edited by K. Itô and J.V. Prokhorov. VIII, 747 pages. 1983.

Vol. 1022: G. Gentili, S. Salamon and J.-P. Vigué. Geometry Seminar "Luigi Bianchi", 1982. Edited by E. Vesentini. VI, 177 pages. 1983.

Vol. 1023: S. McAdam, Asymptotic Prime Divisors. IX, 118 pages. 1983.

Vol. 1024: Lie Group Representations I. Proceedings, 1982–1983. Edited by R. Herb, R. Lipsman and J. Rosenberg. IX, 369 pages. 1983.

Vol. 1025: D. Tanré, Homotopie Rationnelle: Modèles de Chen, Quillen, Sullivan. X, 211 pages. 1983.

Vol. 1026: W. Plesken, Group Rings of Finite Groups Over p-adic Integers. V, 151 pages. 1983.

Vol. 1027: M. Hasumi, Hardy Classes on Infinitely Connected Riemann Surfaces. XII, 280 pages. 1983.

Vol. 1028: Séminaire d'Analyse P. Lelong – P. Dolbeault – H. Skoda. Années 1981/1983. Edité par P. Lelong, P. Dolbeault et H. Skoda. VIII, 328 pages. 1983.

Vol. 1029: Séminaire d'Algèbre Paul Dubreil et Marie-Paule Malliavin. Proceedings, 1982. Edité par M.-P. Malliavin. V, 339 pages. 1983.

Vol. 1030: U. Christian, Selberg's Zeta-, L-, and Eisensteinseries. XII, 196 pages. 1983.

Vol. 1031: Dynamics and Processes. Proceedings, 1981. Edited by Ph. Blanchard and L. Streit. IX, 213 pages. 1983.

Vol. 1032: Ordinary Differential Equations and Operators. Proceedings, 1982. Edited by W. N. Everitt and R. T. Lewis. XV, 521 pages. 1983.

Vol. 1033: Measure Theory and its Applications. Proceedings, 1982. Edited by J. M. Belley, J. Dubois and P. Morales. XV, 317 pages. 1983.

Vol. 1034: J. Musielak, Orlicz Spaces and Modular Spaces. V, 222 pages. 1983.